Chemistry

Teachers' Guide II

Topics 13 to 19

14 082654 8
Nuffield Advanced Science

Project team

E. H. Coulson, formerly of County High School, Braintree (organizer)
A. W. B. Aylmer-Kelly, formerly of Royal Grammar School, Worcester
Dr E. Glynn, formerly of Croydon Technical College
H. R. Jones, formerly of Carlett Park College of Further Education
A. J. Malpas, formerly of Highgate School
Dr A. L. Mansell, formerly of Hatfield College of Technology
J. C. Mathews, King Edward VII School, Lytham
Dr G. Van Praagh, formerly of Christ's Hospital
J. G. Raitt, formerly of Department of Education, University of Cambridge
B. J. Stokes, King's College School, Wimbledon
R. Tremlett, College of St Mark and St John
M. D. W. Vokins, Clifton College

Chemistry

Teachers' Guide II

Topics 13 to 19

Nuffield Advanced Science
Published for the Nuffield Foundation by Penguin Books

Penguin Books Ltd, Harmondsworth, Middlesex, England
Penguin Books Inc., 7110 Ambassador Road, Baltimore, Md 21207, U.S.A.
Penguin Books Ltd, Ringwood, Victoria, Australia

Filmset in 'Monophoto' Times New Roman by
Keyspools Ltd, Golborne, Lancs
and made and printed in Great Britain by
C. Tinling & Co. Ltd, Prescot & London

Design and art direction by Ivan and Robin Dodd
Illustrations designed and produced by Penguin Education

Contents

Foreword

Sixth form courses in Britain have received more than their fair share of blessing and cursing in the last twenty years: blessing, because their demands, their compass, and their teachers are often of a standard which in other countries would be found in the first year of a longer university course than ours: cursing, because this same fact sets a heavy cloud of university expectation on their horizon (with awkward results for those who finish their education at the age of 18) and limits severely the number of subjects that can be studied in the sixth form.

So advanced work, suitable for students between the ages of 16 and 18, is at the centre of discussions on the curriculum. It need not, of course, be in a 'sixth form' at all, but in an educational institution other than a school. In any case, the emphasis on the requirements of those who will not go to a university or other institute of higher education is increasing, and will probably continue to do so; and the need is for courses which are satisfying and intellectually exciting in themselves – not for courses which are simply passports to further study.

Advanced science courses are therefore both an interesting and a difficult venture. Yet fresh work on advanced science teaching was obviously needed if new approaches to the subject (with all the implications that these have for pupils' interest in learning science and adults' interest in teaching it) were not to fail in their effect. The Trustees of the Nuffield Foundation therefore agreed to support teams, on the same model as had been followed in their other science projects, to produce advanced courses in Physical Science, in Physics, in Chemistry, and in Biological Science. It was realized that the task would be an immense one, partly because of the universities' special interest in the approach and content of these courses, partly because the growing size of sixth forms underlined the point that advanced work was not *solely* a preparation for a degree course, and partly because the blending of Physics and Chemistry in a single advanced Physical Science course was bound to produce problems. Yet, in spite of these pressures, the emphasis here, as in the other Nuffield Science courses, is on learning rather than on being taught, on understanding rather than amassing information, on finding out rather than on being told: this emphasis is central to all worthwhile attempts at curriculum renewal.

If these advanced courses meet with the success and appreciation which I believe they deserve, then the credit will belong to a large number of people, in the teams and the consultative committees, in schools and universities, in authorities and councils, and associations and boards: once again it has been the Foundation's privilege to provide a point at which the imaginative and helpful efforts of many could come together.

Brian Young
Director of the Nuffield Foundation

Contributors

Many people have contributed to this book. Final decisions on the content and treatment of the teaching scheme with which it is concerned were made by members of the Headquarters Team, who were also responsible for assembling and writing the material for the several draft versions that were used in school trials of the course. During this exercise much valuable help and advice was generously given by teachers in schools, and in universities and other institutions of higher education. In particular, the comments and suggestions of teachers taking part in the school trials have made a vital contribution to the final form of the published material. Acknowledgement for this assistance is given at the end of *Teachers' Guide II*. The editing of the final manuscript was carried out by B. J. Stokes.

At a time when the Headquarters Team was small in number and under heavy pressure, the following teachers undertook between them the task of developing experimental investigations and planning the theoretical treatment for the first drafts of Topics 12, 14, 15, 16, 17, 18, and 19: W. H. Francis and G. R. Grace (Apsley Grammar School, Hemel Hempstead), K. C. Horncastle (Exeter School), D. R. P. Jolly (Berkhamsted School), Dr R. Kempa (College of St Mark and St John), J. A. Kent (Southfield School, Oxford), P. Meredith (Exeter School), A. G. Moll (Plymouth College), A. B. Newall (the Grammar School for Boys, Cambridge), M. Pailthorpe (Harrow School), Miss K. Rennert (Cheney Girls' School, Oxford), T. A. G. Silk (Blundell's School), Dr R. C. Whitfield (Department of Education, University of Cambridge). It is fitting that their contribution should be acknowledged here. Thanks are also due to D. R. Browning, for advice and assistance with regard to safety measures in laboratory work.

E. H. Coulson

Topic 13
Carbon chemistry, part 2

Objectives

1   In general the objectives of Topic 9, Carbon chemistry, part 1, are to be developed and extended.

2   To consider the chemistry of unsaturated compounds, alkenes, arenes and carbonyl compounds, and also the chemistry of the amino group, $-NH_2$.

3   To develop an appreciation of the need for practical skill, and the importance of yield, by attempting a two or three stage organic synthesis.

Content

13.1   Unsaturated compounds: the alkenes.
13.2   Unsaturated compounds: the carbonyl group in aldehydes and ketones.
13.3   Unsaturated compounds: the arenes.
13.4   Amines.
13.5   Two different functional groups in the same molecule.
13.6   A problem in synthesis.

Timing

Four and a half weeks.

13.1 Unsaturated compounds: the alkenes

Objectives
1 To establish the industrial importance of alkenes in the petrochemical industry.

2 To examine addition reactions of alkenes further to those considered in 9.2.

Timing
Two periods.

Suggested treatment
In this and subsequent sections of this topic OP transparency number 104 will be useful.

Students should be asked to revise their work on alkenes in section 9.2 before this section is considered. Then they can look at the summary of reactions given in the *Students' Book* and consider if a similar mechanism to that deduced for the addition of bromine to ethylene in section 9.2 is valid for these further reactions. An example of such a mechanism might be

$$CH_3CH{=}CH_2 \xrightarrow{\ H^+\ } CH_3{-}^+CH{-}CH_3$$

$$\downarrow {\scriptstyle HSO_4^-}$$

$$\underset{\displaystyle CH_3{-}CH{-}CH_3}{OSO_3H}$$

Polymerization of ethylene by the high pressure process proceeds by a free radical mechanism. Teachers need not discuss this, but students' attention could be directed to the account given in *The chemist in action* (chapter 4). The nature of Ziegler catalysts is discussed in Topic 18.

Teachers may wish to demonstrate again some of the reactions of alkenes and if styrene is used the polymerization can also be observed.

Experiment 13.1
To demonstrate the polymerization of styrene

For this demonstration the teacher will need:

Glass tubing, approximately 1 m long, sealed at one end
6 ball bearings
Styrene
1,2-dibromoethane
Tin(IV) bromide

Procedure

Dissolve approximately 0.5 g of tin(IV) bromide (as catalyst) in 5 cm^3 of 1,2-dibromoethane (as solvent), decanting any insoluble material. Add 5 cm^3 of styrene and pour the reaction mixture into a 1 m length of glass tubing sealed at one end. Drop in a ball bearing and measure the time of fall through the liquid. Repeat this observation during the next two to three days and note the increase in time of fall as the polystyrene chain length increases.

An alternative polymerization procedure for students is given in Topic 18. Students' appreciation of petrochemicals could be usefully extended by additional reading from, for example, the booklet *Petroleum chemicals* available from BP Chemicals (UK) Limited.

The revision questions at the end of this section could be answered as follows:
1   Oct-1-ene and hydrobromic acid give 2-bromooctane.
2   But-1-ene with carbon monoxide and hydrogen gives pentanal.
3   But-1-ene plus sulphuric acid followed by dilution gives butan-2-ol.
4   Propylene heated under pressure with carbon monoxide and hydrogen produces butanal, which can be reduced by hydrogen and a nickel catalyst to give butan-1-ol.

The non-systematic names ethylene and propylene are used here because of their common usage in industry.

Supporting homework
Revision of work on alkenes in section 9.2.
Preparation of a chart showing the industrial importance of propylene.
Reading about petrochemicals.

Supporting material
BP Chemicals (UK) Limited booklet *Petroleum chemicals*, available from Education Section, Information Division, BP Chemicals (UK) Limited, West Halkin House, West Halkin Street, London SW1.

Summary
From this section students will be expected to know something of the nature of a
C=C double bond, the sort of reactions which alkenes undergo, and what is
meant by addition and polymerization. These ideas will be used in the next section, when the students go on to examine a different group containing a double
bond.

13.2   Unsaturated compounds:
the carbonyl group in aldehydes and ketones

Objectives
1   To build up a body of knowledge about the carbonyl group in aldehydes
and ketones.
2   To compare the properties of the double bonds in C=C and C=O.
3   To introduce students to the identification of organic compounds using
the preparation of crystalline derivatives and the determination of their melting
points.

Timing
Five periods.

Suggested treatment
This section is concerned with establishing the general pattern of behaviour of
aldehydes and ketones. It is not intended that attention should be directed towards the behaviour of individual members, such as the unusual reactions of
formaldehyde.

In this section, the *Students' Book* contains an account of the structure of aldehydes and ketones, and this is immediately followed by experiment 13.2. The
object of this experiment is to show how it is possible to take a compound containing a carbonyl group and find out firstly whether it is an aldehyde or a
ketone, and secondly which aldehyde or ketone it is.

For the solution of the first problem, students are introduced to the use of
Fehling's solution (a mild oxidizing agent) which depends on the greater ease of
oxidation of aldehydes compared to ketones. The oxidation of ketones requires
attack on a CH_2 group and the splitting of a C—C bond. For the second part,
they will prepare a crystalline derivative and take its melting point. The most
suitable derivative is the 2,4-dinitrophenylhydrazone, and the reasons for
choosing such a complicated product are described in the *Students' Book*, together with the equation for the reaction.

If the melting point of the derivative has not given a clear identification of the unknown compound, the experiment can be concluded with a determination of boiling point, so as to have two physical quantities to help in the identification.

Experiment 13.2
Identification of an aldehyde or ketone

Each student or pair of students will need:

> Hirsch filtration apparatus
> Melting point apparatus
> Melting point tubes
> Stand and clamp
> Test-tubes, 150×25 mm and 150×16 mm
> Thermometer, 0–250 °C
> Small beaker containing dibutylphthalate
> Sample of aldehyde or ketone (unknown)
> Access to an aldehyde and a ketone (benzaldehyde is not suitable here)
> Solution of 2,4-dinitrophenylhydrazine (see below)
> Methanol
> Ethanol
> Fehling's solution

Procedure

The *Students' Book* contains full instructions for this experiment, together with a set of data of boiling points and melting points of derivatives of representative aldehydes and ketones.

The technique required for the determination of melting points was discussed in experiment 9.4b. Boiling points can be found by gently boiling 0.5 to 1.0 cm^3 of the liquid, in contact with one or two anti-bumping granules, in a 150×16 mm test-tube, with a thermometer suspended just above the level of the liquid. The best results are obtained if the test-tube is immersed in a beaker containing dibutyl phthalate, which should be heated until the thermometer shows a constant reading and the ring of liquid refluxing in the test-tube is about 1 cm above the bulb of the thermometer.

A solution of 2,4-dinitrophenylhydrazine is prepared as follows. To 2 g of 2,4-dinitrophenylhydrazine add 4 cm^3 of concentrated sulphuric acid, cautiously followed by 30 cm^3 of methanol. Warm, if necessary, until the solid has all dissolved, and then add 10 cm^3 of distilled water.

The *Students' Book* continues with a short list of reactions of carbonyl compounds, and asks the reader to compare and contrast these compounds with the alkenes. This can be discussed in class, and the students can be asked to suggest possible reasons for the differences based on the likely distributions of electrons in the alkene and carbonyl groups.

When discussing this latter point it should be borne in mind that oxygen is a more electronegative element than carbon, and so the oxygen atom will be negatively charged with respect to the carbon atom. This is indicated thus

$$\overset{\delta+ \quad \delta-}{\diagdown\!C\!=\!O}$$

Now, whether the reaction takes place in two stages (like the additions to alkenes) or in one, it is clear that the carbon, being positively charged, will be attacked preferentially by negatively charged groups, while the oxygen atom will be attacked by positively charged groups.

This can be illustrated by considering the reaction with hydrogen cyanide, HCN, in the presence of a little alkali to provide some CN^- ions.

or reaction with alcoholic ammonia

Students are expected to produce the following answers to the revision questions set at the end of this section:

1   Dodecanal reacts with hydrogen and a large excess of ammonia in alcoholic solution in the presence of a nickel catalyst to give dodecylamine.

2   Butanone reacts with hydrogen in the presence of a nickel catalyst to give butan-2-ol.

3   Pentanal reacts with acidified dichromate to give pentanoic acid.

4   Heptanal reacts with hydrogen cyanide followed by hydrolysis with aqueous acid to give 1-hydroxyoctanoic acid.

This section of the *Students' Book* ends with some background reading on diabetes, including details of testing for sugars and acetone. If students are interested in this item, dilute aqueous solutions of glucose and acetone ($1 \, g \, dm^{-3}$) could be tested with the clinical preparations which are available from dispensing chemists. These include Acetest reagent tablets for ketones in urine, Clinitest reagent tablets for glucose in urine, and Clinistix reagent strips (based on an enzyme) also for glucose in urine. Alternatively, the laboratory chemicals required are a finely ground mixture of ammonium sulphate (100 parts) and sodium nitroprusside (1 part), and Benedict's solution. To prepare Benedict's solution dissolve 17.3 g of sodium citrate and 20 g of sodium carbonate deca-hydrate in 60 cm^3 of hot water, and add a solution of 1.73 g of copper(II) sulphate pentahydrate in 10 cm^3 water. Make up to 100 cm^3, cool, and filter if necessary.

Supporting homework
Make a comparative study of the reactions of alkenes and the carbonyl group. Learn the reactions of the carbonyl group given in the *Students' Book*.

Summary
After this section, students should be acquainted with the more important reactions of aldehydes and ketones, and be able to contrast these with each other and with reactions of alkenes.

13.3   Unsaturated compounds: the arenes

Objectives
1   To examine experimentally the bromination and nitration of benzene, phenol, and chlorobenzene.
2   To discuss at an elementary level the process of arene substitution by positively polarized attacking groups.

Timing
About a week.

Suggested treatment
Before starting work on this section students should revise their knowledge of the concepts of delocalization, polarization, and dipole moments introduced in Topic 8. They should also look again at their work in Topic 9 on hydrocarbons (alkenes and arenes), phenol, and chlorobenzene.

In the *Students' Book* there is a brief reminder of the delocalized structure of benzene, that is, ⬡ rather than ⬡ , and also of the substitution, rather than addition, reactions that they met in section 9.2. After reading this introduction it is suggested that students then carry out experiment 13.3a, in which they study the conditions for the bromination of benzene, phenol, and chlorobenzene. These are test-tube scale experiments in which no attempt is made to separate products.

Experiment 13.3a

The bromination of arenes

Each student or pair of students will need:

Test-tubes, 150×25 mm, in rack
Teat pipette
Benzene (in a fume cupboard)
Bromine (in a fume cupboard)
Bromine water
Chlorobenzene
Iron (a tin tack)
Phenol

Procedure

In this experiment, details of which are given in the *Students' Book,* students compare the action of bromine water on benzene and phenol. They will see that the colour of the bromine remains when benzene is used, although much of the bromine transfers to the benzene layer. When phenol is used, they will see the colour disappear, and white crystals of 2,4,6-tribromophenol will form. They then see the conditions necessary for the bromination of benzene, and note that these are more drastic than those required for the bromination of phenol.

Finally the bromination of chlorobenzene is examined. The more drastic conditions mentioned in the *Students' Book* are necessary, and reaction may be fitful.

The ease of bromination will be found to be phenol > benzene > chlorobenzene. A series of questions, together with further information, is then posed about the possible mechanism of the reaction. Students should see that reaction between benzene and iodine monochloride occurs as:

$$\bigcirc + \overset{\delta+}{I} - \overset{\delta-}{Cl} \quad \rightarrow \quad \overset{H}{\underset{+}{\bigcirc}} I + Cl^- \rightarrow \quad \overset{I}{\bigcirc} + H^+$$

in which the attacking atom is $I^{\delta+}$ and the leaving atom is H^+.

It therefore follows that if substituent groups produce negative charge on the benzene ring they should activate it to attack. Conversely positive charge on the benzene ring should deactivate it to attack. The dipole moment and relative ease of bromination of phenol and chlorobenzene are consistent with this hypothesis. The distribution of charge could be indicated as shown in the following formulae:

$$\overset{\delta+}{\bigcirc} - Cl^{\delta-} \qquad \overset{\delta-}{\bigcirc} - OH^{\delta+}$$

The function of the catalysts for bromination is mentioned in the *Students' Book,* and could also be discussed. It is considered that they function by polarizing the bromine molecule, and also by removing the anion produced as part of the attack on the benzene ring. The iron used in experiment 13.3a, for example, functions in the following way. It first combines with bromine to form iron(III) bromide.

$$2Fe + 3Br_2 \rightarrow 2FeBr_3$$

This then induces polarization in other bromine molecules

$$Br_2 + FeBr_3 \rightarrow \overset{\delta+}{Br} - \overset{\delta-}{Br}.FeBr_3$$

Substitution then proceeds as follows.

It should be emphasized that the catalyst is in fact iron(III) bromide, and not iron.

After the discussion of bromination students can continue with experiment 13.3b on the nitration of benzene, phenol, and chlorobenzene. If time permits the products of nitration can be separated by thin layer chromatography and shown to be *ortho-* and *para-* nitro compounds. The relative ease of nitration is phenol > benzene > chlorobenzene.

Experiment 13.3b
The nitration of arenes

Each student or pair of students will need:

> Test-tubes, 150×25 mm, in rack
> Teat pipette
> Beaker, 150 cm^3 (also used as an ice bath)
> Watchglass
> Porous plate (optional)
> Prepared silica gel thin layer plates for chromatography (see below)
> Melting point tubes
> Beaker, 600 cm^3, and cover (for chromatography)
> Benzene
> Chlorobenzene
> Ethanol
> Nitric acid, concentrated
> *o*-Nitrophenol
> *p*-Nitrophenol
> Phenol
> Sodium nitrate, solid
> Sulphuric acid, concentrated
> Sulphuric acid, dilute
> Trichloromethane

Procedure
Full details of the experiment are given in the *Students' Book*.

Students will find that the conditions described as 'more drastic' are necessary for the nitration of benzene, phenol will nitrate in 'mild' conditions and 'even more drastic' conditions are necessary for any nitration of chlorobenzene to be observed.

At the teacher's discretion, safety spectacles and polythene 'disposable' gloves may be worn because of the hazardous nature of some of the chemicals involved. In particular, chloronitrobenzenes should not be allowed to come in contact with the skin, as they cause dermatitis (accompanied by painful irritation).

A suitable prepared thin layer is MN-polygram silica gel $S-HR/UV_{254}$ available from Camlab (Glass) Limited, Cambridge. It costs £2.50 for 50 sheets, 20×5 cm. Sheets can be cut with scissors and each sheet will supply two groups.

If prepared layers are not available, they can be made (see Nuffield Chemistry *Collected Experiments,* experiment E2.20, page 41) but additional time and equipment will be needed.

Students who finish before the others could be asked to find out how many products have been obtained in the nitration of chlorobenzene, using the thin layer chromatographic technique described in the nitration of phenol.

The mechanism of the nitration reaction in sulphuric acid solution might now be discussed. For the reaction to take place both nitric and sulphuric acids need to be present. These react together thus:

$$HNO_3 + 2H_2SO_4 \rightarrow NO_2^+ + H_3O^+ + 2HSO_4^-$$

Evidence for this reaction includes:
   1   The mixed acid conducts electricity, indicating the presence of ions, and in electrolysis a nitrogen compound travels to the cathode, indicating that a cation is formed from the nitric acid.
   2   Spectroscopic evidence indicates the presence of the nitronium ion, NO_2^+.

Kinetic studies suggest that it is the nitronium ion which attacks the benzene ring because the rate of nitration is proportional to the concentration of the nitronium ion.

At this point teachers could introduce information about further compounds to validate the ideas discussed (table 13.3a).

Compound	Rate of nitration (relative to benzene)	Dipole moment (δ^-) ↔
NO_2—⟨◯⟩	10^{-4}	3.9
Cl—◯	3×10^{-2}	1.6
◯	1	0.0
◯—CH_3	25	0.3
CH_3 ◯—CH_3	120	—
◯—OH	10^3	1.6

Table 13.3a
Rates of nitration of substituted arenes

In the time available it is not considered possible to develop arene chemistry further, except to mention the sulphonation reaction (important for the manufacture of detergents, discussed in Topic 18) and the diazonium reaction of aniline which students will meet in the next section.

This means that the intriguing subject of *ortho, para,* or *meta* disubstitution is not considered, nor, with the information available to them, will students be able to solve problems of arene synthesis comparable to the problems proposed in 9.7 for aliphatic compounds. Some teachers may like to spend a little time on these topics and curtail the treatment of a topic elsewhere in the course. Two books are listed under 'Supporting material' which teachers may find helpful if they undertake this alternative work.

Finally, the *Students' Book* contains an account of the manufacture of phenol. This account has been chosen to illustrate the effect which social and financial considerations have on the chemistry of industrial processes, and a useful classroom discussion could be based on this, and any other examples known to the teacher. The emphasis in the discussion should be on the criteria of choice used in industry, rather than on the chemistry of the processes (which students should not be expected to learn).

Supporting homework
Revise Topic 8: delocalization, polarization, dipole moment.
Revise Topic 9: alkenes and arenes, phenol and chlorobenzene.

Supporting material
The following books provide useful background for the teacher.
Sykes, P. (1967) *A guidebook to mechanism in organic chemistry*, 2nd edition, Longman.
Clark, N. G. (1964) *Modern organic chemistry*, Oxford University Press.

Summary
As a result of their work in this section students should know that substitution reactions are important in arene chemistry. They should also have an elementary understanding of the process of arene substitution.

13.4   **Amines**

Objective
To study some carbon compounds with basic properties.

Timing
About half a week

Suggested treatment
In this section the students will investigate the effect of introducing alkyl and aryl groups into the ammonia molecule, by comparing some of the properties of butylamine and aniline with those of ammonia. Instructions are given as experiment 13.4.

Experiment 13.4

What is the effect of putting an alkyl or aryl group into the ammonia molecule?

Each student or pair of students will need:

Test-tubes and rack
Hard-glass watchglass
Teat pipette
3 beakers, 250 cm^3
Hydrochloric acid, dilute
Copper(II) sulphate solution
Sodium hydroxide solution
Butylamine
Aniline
Ammonia solution
Ammonium chloride, solid
Sodium nitrite, solid
2-Naphthol
Thermometer, 0–110 °C
Ice

Procedure

Necessary details are given in the *Students' Book* and the results are discussed below.

In the 'blank' diazotization reaction mixture, a pale yellow crystalline solid will separate on adding the 2-naphthol solution, due to nitrosation of 2-naphthol. The same product may be observed in the butylamine reaction mixture if an excess of sodium nitrite is present. Acidification quickly turns this into a black tar, but this should only be done in a 'throw-away' container.

In discussing the results of the experiments the teacher might like to introduce, and demonstrate, the properties of acetamide as a further structural type.

As a result of their work students will discover that the carbon compounds are weak bases like ammonia (table 13.4).

Compound	Base dissociation constant
	$\dfrac{[RNH_3^+][OH^-]}{[RNH_2]}$ mol dm^{-3}
butylamine	4×10^{-4}
ammonia	1.8×10^{-5}
aniline	3.8×10^{-10}
acetamide	7×10^{-15}

Table 13.4
Dissociation constants of amines

The basic strength is related to the readiness with which the compounds will take up a proton on the unshared pair of nitrogen electrons

$$R \underset{H}{\overset{H}{\diagup}} N: + HOH \rightleftharpoons RNH_3^+ OH^-$$

The polarization of a C—N bond expected from the relative electronegativities of the two atoms is $\overset{\delta+}{C}$—$\overset{\delta-}{N}$. The substitution of an alkyl group into ammonia will therefore make the unshared pair of nitrogen electrons more available, and hence alkylamines are stronger bases than ammonia.

The carbonyl group, on the other hand, will act in the opposite sense, and de-localization effects are also possible.

$$R—\overset{O}{\underset{\|}{C}}—NH_2 \rightleftharpoons R—\overset{O^-}{\underset{|}{C}}=NH_2^+$$

These would inhibit the taking up of a proton. For these reasons acetamide is a very weak base, and it is suggested that a delocalization effect is also responsible for aniline being a weaker base than ammonia.

When amines dissolve in water the interaction taking place is hydrogen bonding (Topic 11, Solvation); to recover the amines from hydrochloric acid the solution should be made alkaline and saturated with salt. When the amines react with copper(II) sulphate solution complex ions are formed.

In the reactions with nitrous acid, ammonia produces nitrogen gas, alkyl-amines produce nitrogen and a mixture of alcohol, alkene, and nitroalkane, but arylamines produce relatively stable diazonium salts.

The nitrosonium ion, NO^+, is considered to be the attacking group in conditions of high acidity and in a sequence of steps the carbonium ion, $R—N \equiv N^+$, results. Unless this ion is stabilized in some way a nitrogen molecule will be formed:

$$RNH_2 \xrightarrow{NO^+} R—N \equiv N^+ \rightarrow R^+ + N_2$$

The new ion R^+ will then take part in a variety of further reactions. In the case of arenes stabilization occurs by delocalizing interaction with the benzene ring. The diazonium salts that are formed are useful reactive intermediates.

After the practical work the *Students' Book* continues with an account of the beginning of the modern organic chemical industry, and mentions the leading role of dyestuffs in this connection. (A coloured photograph of the Victoria penny mauve stamp is shown in plate 4 of *Students' Book I.*)

The *Students' Book* on this occasion does not conclude with a summary of the reactions of amines as they are an end product of synthesis rather than intermediates. Students could however gather together from Topics 9 and 13 the different reactions which produce good yields of amines.

Supporting homework
Reading background information in the *Students' Book*.
Listing the methods of preparing amines.

Summary
Students should now be aware of some properties of primary alkyl and aryl amines, and of their relations with those of ammonia. They should also be familiar with some aspects of the development of the organic chemical industry.

13.5   Two different functional groups in the same molecule

Objectives
1   To implant the idea that functional group properties are broadly additive, by examining some properties of amino acids.
2   To introduce the idea of asymmetry of molecular structure.

Timing
About one week.

Suggested treatment
For this treatment OP transparency number 56 will be useful.

Amino acids are used here as examples of compounds containing two functional groups in the same molecule. They can be introduced to the students as compounds having molecules containing both the $-NH_2$ and $-CO_2H$ groups, and it should be pointed out that, in what follows, attention will be restricted to the main type of amino acid which occurs in nature, namely α-amino acids, of general formula

$$
\begin{array}{c}
H \\
| \\
R-C-CO_2H \\
| \\
NH_2
\end{array}
$$

If R is H, the molecule is that of glycine. Students can begin this section by examining some of the properties of this compound, using experiment 13.5a.

Experiment 13.5a
Properties of glycine

Each student or pair of students will need:

> Test-tubes and rack
> 3 microscope slides
> 3 pieces of filter paper the size of the microscope slides
> 6 lengths of electrical connecting wire, with crocodile clips on one end
> Glycine
> Universal indicator solution
> Dilute acids and alkalis
> Copper(II) sulphate solution
> Ninhydrin spray or solution

access to:

> A source of about 50 V d.c. (an H.T. battery would be best)
> An oven

In the course of this experiment, the students are introduced to *paper electrophoresis* and the use of the ninhydrin reagent for the detection of amino acids. If
there is time, this could be followed up by running several different amino acids
on the same piece of filter paper, both separately and as a mixture, showing that
electrophoresis can be used to separate mixtures in the same way as chromatography.

The main point to be made here, however, is not the use of electrophoresis as a
method of separation, but that the process is possible because of the ionic nature
of the amino acids. Furthermore, because of their unusual character (forming
both negative and positive ions) the direction of travel depends upon the pH of
the solution. The Nuffield film loop 'Applications of paper chromatography'
could be shown here, as a contrast to the electrophoresis technique.

Separation by chromatography is carried out experimentally later in the course
(Topic 18).

If R is not hydrogen (or NH_2 or CO_2H) but is some other group of atoms, then
the structure has four different groups attached to the same carbon atom, and
the molecule is therefore asymmetrical.

Figure 13.5a
Space-filling models of L- and D-alanine.

The students can each be asked to construct a model of the molecule of such an amino acid: alanine, in which R is CH_3 (see figure 13.5a), would be a suitable example. They should then compare their models with each other. The two stereo-isomers will probably be seen; if by chance all the class have produced one isomer, the teacher should construct the other. The class could be asked to describe the relationship of the isomers to each other, and to suggest any likely consequences of this that might occur to them.

It may well be necessary to stimulate the students' thinking here by some leading questions, and to simplify the molecular structure by the use of a pair of models, each made from four different coloured spheres grouped tetrahedrally around one central sphere.

One is here trying to establish the object–mirror image relationship of the two structures as a criterion of asymmetry (which can be followed by pointing out the existence of four different groups attached to one C atom) and to recall the use of polarized light to detect asymmetry of crystal structure. Will it detect asymmetry of molecular structure?

If we are to test this by experiment, we shall need a sample of one pure enantiomorph. It should be pointed out that ordinary chemical synthesis yields both enantiomorphs in equal amounts, and so any optical activity is likely to be 'cancelled out'. Furthermore, since both enantiomorphs have the same structural formula, that is, they have the same atoms joined together in the same order, they have the same chemical properties and the same physical properties (such as boiling point, melting point, and solubility) too. They are thus likely to be very difficult to separate.

Biological systems however often synthesize only one of a pair of enantiomorphs. Amino acids obtained from natural sources, for example from protein, are therefore suitable compounds on which to carry out this experiment. In addition to finding out if asymmetry of molecular structure has an effect on polarized light, by doing this experiment students will also find the effects of concentration of solution, and length of column of liquid, on the extent of the rotation of the plane of polarized light.

Experiment 13.5b
　　　To find out if polarized light is affected by
　　　asymmetric isomers of compounds

The two amino acids chosen for this experiment are:

$$\underset{\underset{NH_2}{|}}{HO_2C-\overset{\overset{H}{|}}{C}-CH_2-S-S-CH_2-\underset{\underset{NH_2}{|}}{\overset{\overset{H}{|}}{C}}-CO_2H} \qquad \text{L}(-)\text{-cystine}$$

and

$$HO_2C-CH_2-CH_2-\underset{\underset{NH_2}{|}}{\overset{\overset{H}{|}}{C}}-CO_2H \qquad \text{L}(+)\text{-glutamic acid}$$

These compounds have been selected because
　　　they have a relatively large effect on polarized light [L(−)-cystine has specific rotation −223° and that of L(+)-glutamic acid is +30.4°];
　　　they are also relatively inexpensive.
This information should not be given to the students before they do the experiment!

Each student or pair of students will need:
　　　Two specimen tubes, 75 × 12 mm
　　　Thin glass stirring rod
　　　L(+)-glutamic acid, 5 g
　　　L(−)-cystine, 1 g
　　　Approximately M hydrochloric acid, 25 cm³

In addition, a polarimeter should be set up in a shaded part of the laboratory; one of these instruments will serve for about ten students. The diagram shows a suitable form of instrument and how it is best employed (figure 13.5b).

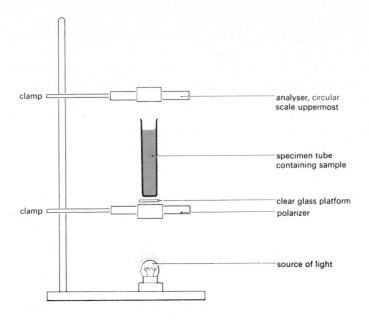

clamp — analyser, circular scale uppermost

specimen tube containing sample

clear glass platform

clamp — polarizer

source of light

Figure 13.5b

A simple polarimeter. A red or blue filter on the glass plate increases sensitivity.

Full details of the experiment are given in the *Students' Book*. When the experiment is completed there is a good opportunity to discuss optical isomerism, taking account of the conclusions drawn from the experiment. The teacher should aim to cover the following points.

Experimental findings

1   Compounds with asymmetrical molecules rotate the plane of polarization of plane-polarized light when in solution. Those compounds which were found to affect polarized light in the solid state when investigated in Topic 8 did so because of asymmetry of *crystal* structure. That this is not due to any asymmetry of *molecular* structure can be seen by dissolving a few optically-active crystals (such as those of sodium chlorate) and examining the solution, which has no optical activity. This should certainly be attempted whilst the polarimeter is set up, if there appears to be any doubt in the students' minds.

2   The angle of rotation is different for different substances.

3   For a given substance the angle of rotation depends upon

   a the concentration of the solution, and

   b the length of column of liquid.

(It also depends upon the wavelength of the light, and the temperature, but these are not investigated.)

Definitions

Students should realize the need for specificity when quoting angles of rotation, and should be able to work out specific rotations.

1   Specific rotation is the angle through which the plane of polarized light is rotated by a 10 cm column of liquid containing a concentration of 1 g of optically active compound per cm^3 of solvent. It is usually measured using light of the wavelength of the sodium D line, at 20 °C, and in this case is denoted by the symbol $[\alpha]_D^{20}$. The specific rotation is calculated from a measured angle α_D^{20} using the expression

$$[\alpha]_D^{20} = \frac{\alpha_D^{20}}{l \times d}$$

where l is the length of the experimental column of liquid measured in decimetres and d is the number of grammes of compound in each cm^3 of solution.

2   Rotation of the plane of polarization in the clockwise sense as viewed by an observer looking towards the source of light is given a + sign.

Example. $[\alpha]^{20}$ for (+)-glutamic acid is +30.4°.

Compounds having this property are described as *dextrorotatary* and have a prefix (+) to their names.

3   Rotation of the plane of polarization in the anticlockwise sense as viewed by an observer looking towards the source of light is given a − sign.

Example. $[\alpha]_D^{20}$ for (−)-cystine is −223°.

Compounds having this property are described as *laevorotatory* and have a prefix (−) to their names.

4   Molecules whose structures are related to one another, as are object and mirror image, produce the same amount of rotation but in opposite senses. Pairs of substances having molecules related in this way are known as enantiomorphs. One structure is called the L-structure and the other is called the D-structure. The D and L prefixes to the names of these compounds have structural significance only, and have nothing to do with the sense of rotation (they do *not*, for example, stand for dextrorotatory and laevorotatory). The structures they represent are as now shown.

L amino acids   D amino acids

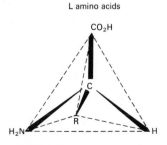 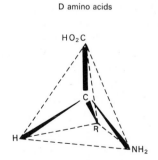

Figure 13.5c
Configuration of amino acids.

Teachers might like to note that the naturally occurring L amino acids can be identified (in diagrams) by the 'corn' law for their substituent groups when viewed along the C^α—H axis, as shown in figure 13.5d.

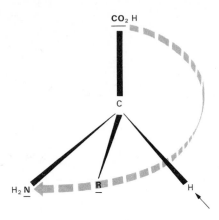

Figure 13.5d

The full name of an optically-active compound thus includes D or L to indicate which structure, and $(+)$ or $(-)$ to indicate which sense of rotation.

Resolution
Resolution is tedious but possible; resolution by compound formation could be discussed and Pasteur's researches described.

Supporting Homework
1   Taking home two pieces of polaroid and a specimen tube (or the Student Polarimeter if allowed) and using these, in conjunction with a torch as a source of light, to find out if any household substances have asymmetrical molecules. Start with strong sugar solution. Is it possible to distinguish between oil of turpentine ('turps') and white spirit ('turps substitute') in this way? (These are both liquids and of course do not need dissolving in anything.)

(*Note to teachers*. 'Genuine' turpentine consists largely of α-pinene. Both $(+)$ and $(-)$ forms are naturally occurring, but some samples of turpentine are optically active. As an alternative students could be asked if they could distinguish between natural and synthetic camphor if samples of either or both were available. It is not intended here that they should learn the names and formulae of these carbon compounds, but merely the use of polarized light as an analytical tool.)

2   Predicting the likely course of the reactions of L(+)-lactic acid with
a sodium,
b sodium hydroxide, anɑ
c phosphorus trichloride.
Are the products likely to be optically active?
3   Making a list of the substances met in Topics 9 and 13 which have asymmetric molecules.

Summary

In this section the students are introduced experimentally to the basic ideas of stereochemistry. Having completed the section they should know:
1   The various factors upon which the angle of rotation depends, and how to calculate the specific rotation of an optically-active substance.
2   The use of D, L, (+) and (−) in the nomenclature of optically-active substances, and the meaning of enantiomorph.
3   One method of resolution of enantiomorphs.

They will have also seen that functional group properties are broadly additive in the case of one class of compound, and should be able to make predictions of simple reactions of other classes of compounds having more than one functional group in each molecule.

13.6   **A problem in synthesis**

Objectives

1   To provide an opportunity for students to use some of their knowledge of organic chemistry and practical technique.
2   To gain experience of the requirements for successful multi-stage organic preparations.

Timing

One week of practical work.

Suggested treatment

As the work involves some forward planning, and could provide a good means of revising the students' knowledge, this section might conveniently be taken later in the course. It would be advantageous if arrangements could be made for at least one whole morning or afternoon to be spent on the practical work, as a great deal more of this type of work can be done in, say, three consecutive hours, than in the same time split up over several days.

In section 9.7 students have already been introduced to problems in organic synthesis so this section affords an opportunity to test practically the ideas introduced there. The *Students' Book* suggests that practical books should be consulted for the students to plan a synthesis scheme of their own choice. It is felt that this is a valuable activity but teachers may have strong reservations about students actually executing their own schemes.

Due to inexperience, many schemes proposed by students may not be feasible and the successful execution of a scheme is as important as the planning. For this reason details of six schemes are given below which might be offered to students who in the teacher's estimation plan unrealistic or unrewarding schemes. The six schemes offered have a bias towards processes of industrial interest. It is hoped, for example, that the production of a dyestuff culminating in its application to some cloth will add more to the student's interest in his work than producing an obscure compound in a specimen tube.

Detailed descriptions of practical technique (suction filtration, drying liquids, setting up apparatus) are not given, as it is felt that these are much better conveyed by a teacher's demonstration which can be related to the particular laboratory facilities available.

Teachers will appreciate that these schemes are examples only, and it is not intended that students should learn the details of these syntheses. The purpose of this section of the topic is that the students shall gain experience of practical work of a preparative nature, and find out something of the problems involved.

Acknowledgement. Some of the schemes offered are based on the work of Dr S. Coffey, ICI Dyestuffs Division.

Reaction Scheme One   Preparation of adipic acid and nylon

The manufacturers of nylon are highly competitive and the cost and yield in each stage of the manufacturing process is subject to rigorous control. Cyclohexane, obtained by hydrogenation of benzene, is first oxidized by air to a mixture of cyclohexanol and cyclohexanone. The mixture is then further oxidized by nitric acid to adipic acid. 1,6-diaminohexane (obtained from adipic acid via adiponitrile) is reacted with the adipic acid to give nylon. In this reaction scheme different laboratory routes to adipic acid can be compared for cost and yield.

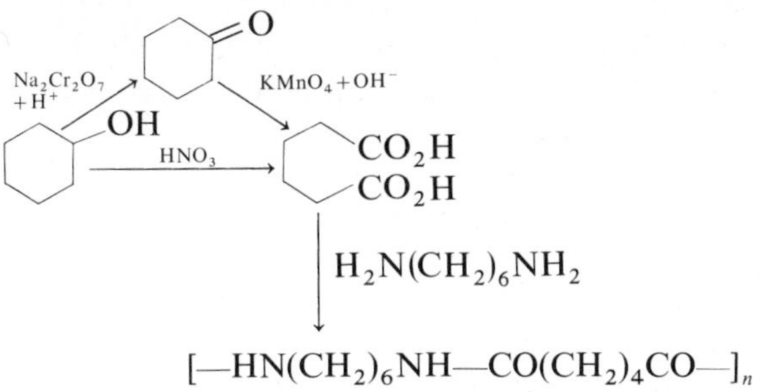

Chemicals required:

Cyclohexanol
Sodium dichromate dihydrate
Glacial acetic acid
Potassium permanganate
9M sulphuric acid (50 per cent by volume)
2M sodium hydroxide solution
1,6-diaminohexane
Nitrogen gas

Preparation of cyclohexanone

Dissolve 30 g sodium dichromate dihydrate in 50 cm^3 glacial acetic acid by warming in a 200 cm^3 conical flask (this will take some time and stirring will be necessary). Cool to 15 °C and then add the solution to a mixture of 32 cm^3 (0.3 mole) cyclohexanol and 25 cm^3 glacial acetic acid cooled in a second flask in ice to below 15 °C, and note the temperature carefully. The reaction soon commences and the temperature rises slowly at first and then rapidly. As the temperature approaches 60 °C cool the flask in ice water, shaking it gently from time to time, *just sufficiently* to keep the temperature from rising further, for about fifteen minutes. After this the temperature will keep up for a time and then fall of its own accord without cooling. The oxidation is now complete, and the solution deep green to dark brown in colour. After standing for a further ten minutes, transfer to a distillation apparatus, wash out the original flask with 100 cm^3 water into the distillation flask, and distil in steam as long as any oil passes over in the distillate.

When the oily drops have ceased, and the distillate comes over clear, continue the distillation until a further 25 cm^3 have been collected. Cyclohexanone is appreciably soluble in water, and this further distillation will ensure that the rest of the product is collected.

Now neutralize the acid distillate with 2M sodium hydroxide solution, add salt up to 20 per cent by weight of the total distillate to salt out the cyclohexanone, separate, dry over anhydrous sodium sulphate, and distil the cyclohexanone, which distils at 157 °C; the yield should be about 15 cm^3.

(To improve the yield the aqueous salted distillate can be extracted with about 30 cm^3 of ether. The ether extract is added to the cyclohexanone, dried out and distilled with suitable precautions against an ether fire.)

Oxidation of cyclohexane derivatives to adipic acid

1   *Direct oxidation of cyclohexanol*
Caution. It is not safe to carry out this experiment with quantities greater than those specified, and it is better performed as a teacher demonstration.

Place 8 cm^3 of concentrated nitric acid in a 250 cm^3 beaker *in a fume cupboard*. The large beaker is necessary because of the vigour of the reaction. Add 2.1 cm^3 (0.02 mole) of cyclohexanol and withdraw your hand at once. When the reaction subsides, cool the beaker under the tap.

Filter off the crude adipic acid at the pump and recrystallize from hot 2M nitric acid, adding a little activated charcoal to remove coloured impurities. Filter free of charcoal and cool the filtrate in a refrigerator.

Collect the recrystallized product by suction filtration, wash free of nitric acid with a little chilled water, and dry on a water bath.

Pure adipic acid is a white crystalline solid, melting point 152–153 °C

2   *Oxidation of cyclohexanone*
Dissolve 30.5 g of potassium permanganate in 500 cm^3 of water and stir in 10.5 cm^3 of cyclohexanone (0.1 mole). Adjust the temperature of the mixture to 30 °C and then add 10 cm^3 of 2M sodium hydroxide, mixing well. The temperature will slowly rise to about 45 °C but the reaction mixture should now be allowed to stand overnight without further attention.

If potassium permanganate is still present the next day add small quantities of sodium hydrogen sulphite to destroy the remaining purple colour. Filter the mixture by suction using a large funnel, washing the precipitate of manganese dioxide with portions of water. Evaporate the combined filtrate and washings to about 70 cm^3, adding anti-bumping granules as a precaution, and filter again if necessary to obtain a clear colourless solution.

Adjust the hot solution to pH 1 (as judged by a suitable indicator paper) by addition of concentrated hydrochloric acid. Then add a 10 cm^3 excess of acid. Allow the solution to cool to room temperature when crystals of adipic acid will appear and may be collected by suction filtration. The yield of adipic acid, melting point 152–153 °C, is about 60 per cent.

Preparation of a polyamide (nylon)

Make separate solutions in methanol of 2 g of adipic acid and 2.5 g 1,6-diaminohexane, mix, cool, and filter off the solid nylon 'salt' which forms. The nylon 'salt' may take some time to form.

Place the dry nylon 'salt' in a wide test-tube clamped in an oil bath. Raise the temperature while passing a slow stream of nitrogen into the test-tube. When 200 °C has been reached, lightly stopper the test-tube and maintain at 200 to 250 °C for a few hours.

The molten mass slowly stiffens as polymerization takes place, and fibres can be drawn out on the end of a glass rod. A brittle fibre is obtained but if the degree of polymerization is correct it should be possible to pull out the fibre ('cold drawing') and obtain a strong flexible fibre.

Reaction Scheme Two   Preparation of an azo dyestuff from benzene

'Nitraniline Orange' is an example of an *azoic* dye. Azoic dyes are water insoluble dyes which are produced directly on the fabric. Thus for the application of Nitraniline Orange the cotton to be dyed is treated with an alkaline solution of 2-naphthol, dried, and then immersed in a bath of diazotized *meta*-nitroaniline. The two components then react to produce the dyestuff.

Nitraniline Orange is one of the oldest azoic dyestuffs and has now been superseded by superior compounds.

The production of dinitrobenzene could be performed in one reaction but it is not economic as much more sulphuric acid has to be used.

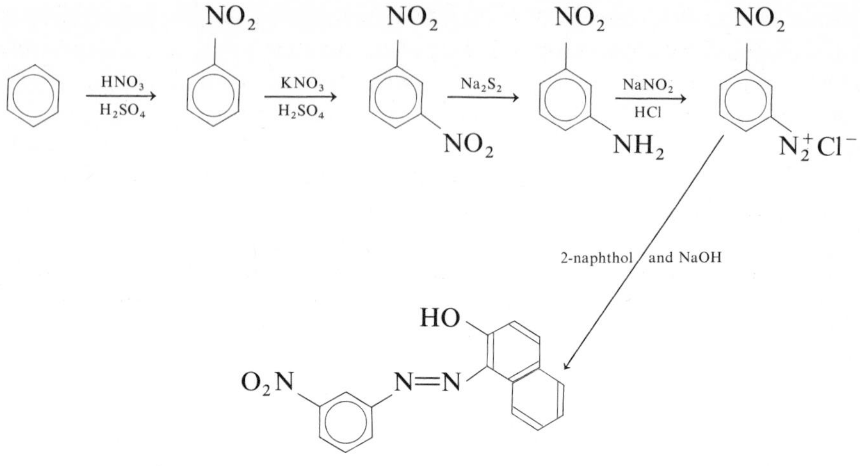

Chemicals required (*other than common laboratory reagents*):

> Benzene
> Potassium nitrate
> Sodium sulphide
> Sulphur
> 2-Naphthol
> Cotton fabric (well washed)
> Sodium nitrite
> Sodium acetate

Preparation of nitrobenzene

Caution. Nitrobenzene is toxic: take care not to breathe the vapour or allow the liquid to come in contact with the skin.

Place 10.5 cm^3 of concentrated nitric acid in a 50 cm^3 flask, and add 12 cm^3 of concentrated sulphuric acid, mixing it in a little at a time. Now, stirring with a thermometer, add 9 cm^3 (0.1 mole) of benzene in portions of about 2 cm^3. Shake the mixture between each addition. Make sure the temperature does not rise above 60 °C or it can become uncontrollable. But do not cool the mixture too much or the reaction may not start properly.

When all the benzene has been added, attach a reflux water condenser and heat the mixture on a water bath at 60 °C for thirty minutes. Shake the mixture occasionally during the refluxing. At the end of the refluxing, cool the flask.

Now pour the mixture, with stirring, into 100 cm^3 of water in a beaker. Decant off the upper aqueous layer as far as possible, add more water to wash the nitrobenzene from most of the residual acid, and remove the washings by decanting. Transfer the residual liquid to a separating funnel and add 20 cm^3 of sodium carbonate solution. Shake gently, frequently releasing the pressure. Allow to stand, then *decant* the upper, aqueous layer.

Repeat the process until no carbon dioxide is formed on adding the carbonate; then transfer the lower, nitrobenzene, layer to a small conical flask and dry it with anhydrous sodium sulphate, shaking occasionally until the liquid is clear.

Set up an apparatus for distillation using an air condenser, and distil the nitrobenzene, collecting the fraction boiling at 207–211 °C. Alternatively the product can be used directly without distillation for the preparation of *meta*-dinitrobenzene.

Preparation of *meta*-dinitrobenzene
Caution. Nitrobenzene is toxic. Take care not to breathe the vapour or allow the liquid to come in contact with the skin.

Place 8 g of potassium nitrate and 16 cm^3 of concentrated sulphuric acid in a 50 cm^3 flask. Add 5 cm^3 (0.05 mole) of nitrobenzene and shake gently. Fit the flask with a reflux air condenser and set up the apparatus in a fume cupboard. Heat the mixture gently to boiling, then reflux on a sand bath for about fifteen minutes with occasional shaking. If there is any sign of the mixture darkening, raise the flask from the sand bath and remove the flame for a short while.

Now allow the flask to cool to about 50 °C (the temperature at which it can be handled without discomfort), remove the reflux condenser and pour the mixture, with stirring, into 50 cm^3 of water in a beaker. The crude dinitrobenzene should solidify.

Filter off the solid by suction filtration, and wash thoroughly with water to remove all traces of acid.

Recrystallize the solid from ethanol. Leave it to crystallize in an ice bath or a refrigerator because *m*-dinitrobenzene is quite soluble in ethanol at room temperature. Filter off the crystals and dry them at room temperature.

A further crop of less pure crystals can be obtained from the filtrate by adding water until a precipitate is formed. Add enough water so that the precipitate just dissolves at boiling point. Cool again in an ice bath or a refrigerator, and filter and dry as before.

The yield is about 60 per cent of pale yellow crystals, melting point 90 °C.

Preparation of *meta*-nitroaniline

Prepare a solution of sodium disulphide by adding 2.1 g of sulphur ('flowers') to a solution of 8 g of sodium sulphide enneahydrate (Na_2S, $9H_2O$) in 30 cm^3 of water, and boiling the mixture until a clear orange solution is obtained. Transfer the cooled solution to a dropping funnel.

Into a suitable flask put 5 g (0.03 mole) of *m*-dinitrobenzene, 100 cm^3 of water, and 50 cm^3 of ethanol. Heat the solution to boiling, and allow the sodium disulphide solution to drop in slowly. The reaction is exothermic, but a little heating may be needed to keep the solution boiling. The reaction mixture should eventually go an orange-brown colour. When all the sodium disulphide solution has been added, boil the solution for a further five minutes.

Filter the hot solution through a hot Buchner funnel to remove solid impurities, then cool the filtrate to room temperature and finish in a refrigerator or an ice bath. Filter off the crystals, wash with water, and drain dry.

Recrystallize the solid from aqueous ethanol.

The yield should be about 75 per cent. The product is a yellow crystalline solid, melting point 114 °C.

Preparation of the Azoic Dye

1   *Treating the fabric*

Dissolve 2 g of 2-naphthol in 50 cm^3 of boiling 2M sodium hydroxide; then add 200 cm^3 of cold water and 1 cm^3 of liquid detergent. Work a 2 g piece of undyed well-washed cotton fabric in the lukewarm liquid for five minutes, wring out, and dry with hot air. Wear protective gloves when handling the fabric. When dry the treated fabric should be developed in the dye bath without delay.

2   *Preparation of the dye bath*

Dissolve 3 g (0.02 mole) of *meta*-nitroaniline in 30 cm^3 of boiling 2M hydro-chloric acid. Add the solution to 50 cm^3 of cold water, stirring well. The chloride will separate as fine crystals. Cool in an ice bath to between 5 and 10 °C and add slowly a similarly chilled solution of 2 g of sodium nitrite in 10 cm^3 of water. Stir well during the addition and ensure that the temperature does not rise above 10 °C. Allow the solution to stand for ten minutes, and then add a solution of 4 g of hydrated sodium acetate in 40 cm^3 of 2M sodium hydroxide.

The prepared cotton fabric is dyed by working in this dye bath for a minute, wringing out and working for a further minute. Finally rinse the dyed cotton and wash it in hot soapy water. Wear protective gloves when handling the fabric.

Reaction Scheme Three   An azo dyestuff for rayon from aniline
Acetate rayon does not contain reactive groups as wool does which help fix dyestuffs to the fabric. It is believed that acetate rayon is dyed by forming a 'solid solution' with dyes (in the same way as ether can extract organic materials from water). The chemicals used in this reaction scheme are those used in industry, where cost is an important factor in the choice of compounds.

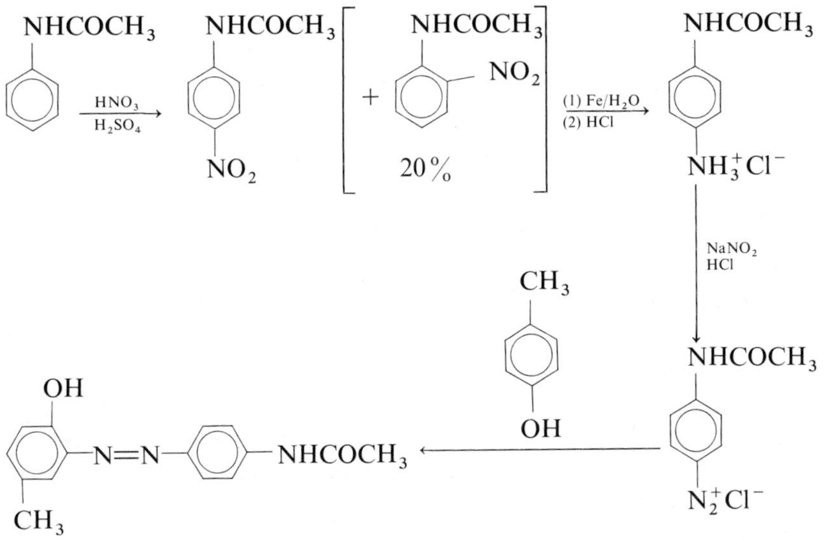

Chemicals required (other than common laboratory reagents) :

> Acetanilide
> Iron filings
> Magnesium oxide
> (Sodium sulphide)
> Sodium nitrite
> *p*-Cresol (*p*-hydroxytoluene)
> Acetate rayon fabric

Preparation of *p*-nitroacetanilide

Into a 250 cm^3 beaker fitted with a stirrer and a cooling bath, introduce 10 cm^3 of concentrated sulphuric acid and stir while cooling in an ice bath. Now add, gradually, 5 g powdered acetanilide, taking care that the temperature never exceeds 25 °C. Stir until dissolved. Cool the mixture in ice and salt to 0 °C, and add slowly, drop by drop, from a dropping funnel to the well stirred and cooled mixture, a previously prepared mixture of 2.5 cm³ concentrated nitric acid with 2.5 cm³ concentrated sulphuric acid. Take care to keep the reaction mixture below 5 °C and as near 0 °C as possible. When all the nitrating acid has been added, keep stirring the mixture, remove the cooling bath and allow it to stand for half an hour. Then pour the reaction mixture with stirring onto 50 g of ice mixed with 50 g of water. Filter the *p*-nitroacetanilide by suction filtration, wash free of acid with ice-cold water and press the filter cake as dry as possible. This wet filter cake can be used in the following experiments. The more soluble *o*-nitroacetanilide is washed away.

A specimen can be dried and recrystallized from ethanol to give *p*-nitroacetanilide as colourless crystals, melting point 214 °C.

Preparation of *p*-aminoacetanilide hydrochloride

Fit up a 250 cm³ flask with a stirrer and a wide-bore reflux condenser, and into it put 8 g of grease-free iron turnings or filings and 50 cm³ of water. Now add 0.5 cm³ of glacial acetic acid and heat the mixture to nearly boiling (95 °C) with stirring. Some hydrogen will be evolved and the iron will become etched. Now add the *p*-nitroacetanilide paste from the previous experiment, in small portions at a time, at such a rate as to keep the stirred reaction mixture near the boil without application of heat. When all the nitro compound has been added, heat the stirred mixture to the boil for half an hour, at which time reduction should be complete.

p-Aminoacetanilide is unstable in alkaline solution and the next stage of filtration and washing should be carried out as rapidly as possible. Make the hot mixture just alkaline with about 1 g of magnesium oxide and test the solution for freedom from iron by spotting on filter paper with sodium sulphide; a black spot means that insufficient magnesium oxide has been added. When all the iron is precipitated, add 0.5 g of sodium sulphite in 1 cm³ of water as a preservative against oxidation, and filter off from unchanged iron and iron oxide on a large suction filter using a large pre-heated funnel and filter flask. Wash the filter cake well by stirring with hot water (60 cm³) and combine the washings and filtrate, cool to 60 °C and add hydrochloric acid (about 20 cm³ of 2M) until a blue colour is produced on Congo Red paper. This filtrate may, if wished, be treated with concentrated hydrochloric acid and diazotized without separation, according to the instructions given in the next section.

With the temperature still at 60 °C, measure the volume of solution and add sufficient salt (NaCl) to make 26 per cent brine. Again test for acidity, re-adjusting for any alkalinity introduced with the salt. Cool to below 20 °C, stirring, filter off the *p*-aminoacetanilide hydrochloride, press well and dry in a desiccator. The product consists of p-aminoacetanilide hydrochloride (50 to 60 per cent) together with solid sodium chloride.

The amount of p-aminoacetanilide hydrochloride in the product may be determined by titration with 0.5M sodium nitrite solution when the amine is converted quantitatively into a diazonium compound.

Weigh accurately about 1 g of the sample and add water at 35 °C until the sample dissolves, dilute with cold water to about 200 cm³ and add 5 cm³ of concentrated hydrochloric acid. Titrate with 0.5M sodium nitrite solution until an immediate faint blue colour is obtained on starch iodide paper five minutes after the last addition of nitrite, added 0.1 cm³ at a time towards the end of the titration.

Per cent p-aminoacetanilide hydrochloride (m.w. 186.7) $= \dfrac{\text{titre} \times 186.7 \times 100}{\text{weight taken} \times 2000}$

$$= \dfrac{\text{titre}}{\text{weight taken}} \times 9.3$$

Preparation of 'Dispersol' Fast Yellow G

The preparation consists of three stages: (a) the preparation of the coupling component, an alkaline solution of p-cresol; (b) the preparation of a solution of the diazonium salt or diazo-component; and (c) the coupling to make the dyestuff.

Prepare a solution of 2.8 g p-cresol in 12.5 cm³ of 2M sodium hydroxide and 80 cm³ of water by stirring the components in a 500 cm³ beaker, and then add 2.7 g anhydrous sodium carbonate and stir until dissolved.

Prepare the diazo-component as follows. Into a 500 cm³ beaker fitted with a mechanical stirrer, put p-aminoacetanilide hydrochloride (equivalent to 4.7 g 100 per cent), 25 cm³ of water, and 18.5 cm³ of 2M hydrochloric acid. Stir and cool to 0 °C in a cooling bath. Now, keeping the temperature between 0 °C and 5 °C, add drop by drop from a burette, with stirring, over about five minutes, a 10 per cent solution of sodium nitrite in water until the starch-iodide paper test shows a slight excess of nitrous acid (about 17.5 cm³ should be required).

For the coupling stage, cool the p-cresolate solution to 10 °C and, with good mechanical stirring, add to it the cold diazonium salt solution drop by drop over fifteen minutes, keeping the temperature of the mixture at 10–15 °C. When addition is complete, stir for an hour if possible, filter off the dyestuff, wash it until it is free of alkali and keep it in paste form (if it is proposed to make a dyeing) or as a dry powder if required as a specimen. The yield of dry powder is 3.5 g.

5 g of acetate rayon fabric can be dyed as follows. Take 0.05 g of dyestuff paste, grind it in a mortar with a couple of drops of a liquid detergent, and then wash it into a beaker with about 250 cm^3 of water and stir the fine suspension. Warm to 50 °C, put in the rayon and raise the temperature slowly to about 90 °C, keeping the fabric or yarn in gentle motion the whole time. After an hour, remove the dyed material, wash in dilute soap solution and dry.

Reaction Scheme Four   Preparation of a constituent of perfumes, 'Essence of Orange Blossoms'

The fragrance of orange blossom has long been admired; in China, orange blossoms are used to scent tea, 'Orange Pekoe', and in the South of France orange trees are grown to make perfume from the blossom rather than for fruit.

The high cost of producing natural perfume has resulted in the production of synthetic materials which are cheap but which will still give satisfaction. 2-Ethoxynaphthalene, or 'Bromelia', can be compounded with other substances to give a cheap substitute for the true essence of orange blossoms, whose odour is attributed to methyl anthranilate. The butyl ether, or 'Fragerol', has a weak strawberry odour.

$$C_2H_5OH \xrightarrow[\text{H}_2\text{SO}_4]{\text{NaBr}} C_2H_5Br \xrightarrow{\substack{\text{sodium salt of}\\\text{2-naphthol}}}$$

Chemicals required (other than common laboratory reagents):

Ethanol
Sodium bromide
2-naphthol

Preparation of bromoethane

Set up the apparatus for distillation with a 100 cm^3 flask. As receiver use a small conical flask containing 10 cm^3 of ice cold water and surround by an ice bath. A delivery tube from the end of the condenser should dip below the surface of the water in the receiver to minimize loss of the low boiling bromoethane.

In the distillation flask place 15 cm^3 of ethanol and 10 cm^3 of water. Add slowly, mixing well, 16 cm^3 of concentrated sulphuric acid. Cool under the tap and add 14 g of sodium bromide. Connect the flask to the condenser without delay.

Heat gently so that oily drops of bromoethane distil over. The contents of the flask may foam, in which case stop heating for a time. When no more bromoethane distils transfer the contents of the receiver to a separating funnel.

Decant off the upper aqueous layer. Wash the bromoethane with an equal volume of 1.5M sodium carbonate and then an equal volume of water, decanting off the aqueous layers. Run the bromoethane through the tap of the separating funnel into a dry flask. Dry it with anhydrous granular calcium chloride and redistil, collecting the fraction having the boiling range 35–40 °C. A very good flow of cold water in the condenser will be required to ensure adequate condensation.

Because of its low boiling point bromoethane should be kept in a well-sealed container.

Preparation of 2-ethoxynaphthalene
Place 3.6 g of 2-naphthol, 1.0 g of solid sodium hydroxide, and 10 cm³ of ethanol in a 100 cm³ flask equipped with a condenser. Reflux the mixture over a low flame for ten minutes until all the sodium hydroxide dissolves; sodium 2-naphthoxide is formed.

Cool the solution to room temperature, and add 3 cm³ of bromoethane; continue refluxing the solution with a low flame for an hour. The severe bumping that develops is due to the sodium bromide that precipitates.

Set the condenser for distillation, and distil 8 to 10 cm³ of volatile liquid still present at the end of the reaction. Discard the distillate, which consists mainly of ethanol and unreacted bromoethane.

Add 50 cm³ of water to the oily residue in the flask and distil the mixture, so that the ether steam distils. Unreacted sodium 2-naphthoxide remains behind as a water-soluble, non-volatile salt. Stop the distillation when no more oily drops occur in the distillate, if necessary adding more water to the flask. If the product solidifies in the condenser during the distillation, temporarily turn off the cooling water of the condenser until the product melts and runs into the receiver flask again.

Cool the distillate in ice water, swirling the receiver flask and scratching the oily product with a glass rod to induce crystallization. When the 2-ethoxynaphthalene has solidified, filter the product with suction and press the filter cake as dry as possible.

Purify the crude 2-ethoxynaphthalene by recrystallizing it from 95 per cent ethanol.

Place the damp crystals in a 150×16 mm test-tube and dissolve them in a minimum volume (1 to 4 cm^3) of ethanol. Only slight warming of the alcohol is required to dissolve the solid so the use of a condenser is not necessary in this instance. Cool the solution in an ice bath, scratching with a glass rod to induce crystallization. Filter the recrystallized product with suction. Dry 2-ethoxy-naphthalene has the melting point 37 °C.

Reaction Scheme Five   Preparation of an insecticide, DDT

DDT was first manufactured on a large scale in 1943 and put to immediate use in the Second World War for killing lice and mosquitoes, thus preventing attacks of typhus and malaria amongst the Allied Troops. Since then the use of DDT has eradicated malaria from countries such as Ceylon and freed people from disease that undermines the ability and will to work.

Limitations have now appeared in the use of insecticides such as DDT because insect species such as the house fly have proved capable of developing resistance to them. Also traces of DDT can now be found throughout the animal world, and it would be wise if in future insecticides were used with more caution until their full effect has been investigated.

Chemicals required:

  Aluminium foil
  Mercury(II) chloride solution (0.1M)
  Benzene (dried)
  Chlorine generator
  Chloral hydrate
  Solid detergent

Preparation of chlorobenzene

Prepare the aluminium amalgam catalyst by just covering about 1 g of thin aluminium foil with 2M sodium hydroxide in a small beaker. Warm until reaction commences. Allow the reaction to continue for two minutes, then pour off the sodium hydroxide and wash the aluminium by decantation, first with water and then with ethanol. Now add sufficient 0.1M mercury(II) chloride solution to cover the aluminium and leave for two minutes. Wash again by decantation, first with water and ethanol and finally with *dry* benzene. The catalyst must be used at once for the preparation of the chlorobenzene.

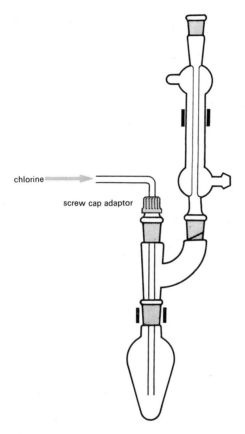

chlorine

screw cap adaptor

Figure 13.6

Set up the apparatus illustrated, or its equivalent, in a fume cupboard, and introduce 22 cm^3 (0.25 mole) *dry* benzene and about 0.1 g of aluminium catalyst. Weigh the flask and contents.

Pass in *dry* chlorine until an increase in weight of about 7 g has occurred (80 per cent theoretically). The reaction is exothermic and hydrogen chloride is evolved. Moderate the reaction by cooling the flask in a water bath.

When the increase in weight has been achieved pour the product into 20 cm^3 of water in a separating funnel. Shake well and decant off the upper water layer. Wash the chlorobenzene with small portions of 2M sodium hydroxide, and water, dry over anhydrous calcium chloride and distil, through a fractionating column if possible, collecting the fraction boiling between 127 and 135 °C. A short length (5–10 cm) of tube packed with metal pot scourer will serve as a fractionating column.

Pure chlorobenzene has boiling point 131–132 °C.

Preparation of DDT
Shake 7 g chloral hydrate with 7 cm^3 concentrated sulphuric acid in a separating funnel (*caution*) until the crystals have liquefied in an endothermic change. This operation may be omitted but some reduction in yield will result. Run off the lower layer of concentrated sulphuric acid and wash the liquid chloral with a second 7 cm^3 portion of acid. To the chloral in the separating funnel add 4 cm^3 of chlorobenzene, 20 cm^3 of concentrated sulphuric acid, and 0.5 g solid synthetic detergent. Shake the mixture to obtain an emulsion of the two layers. Intermittent shaking for two hours will produce a precipitate of DDT but the reaction mixture can be left for a day with occasional shaking.

After the reaction period, pour the mixture with continuous stirring into 100 cm^3 of cold water. Stir well and collect the crude DDT by suction filtration. To remove traces of sulphuric acid shake the crude product with 10 cm^3 of 1.5M sodium carbonate solution, collect by suction filtration, and wash well with several small portions of water.

To obtain a purer product recrystallize the DDT from propan-1-ol at the rate of 5 cm^3 per gramme of material. Pure DDT has a melting point of 108 °C.

Reaction Scheme Six   2,4-Dinitrophenylhydrazine from benzene

In experiment 13.2, 2,4-dinitrophenylhydrazine was used to prepare crystalline derivatives of carbonyl compounds. If pure, the derivatives have sharp melting points and the carbonyl compound can be thus identified.

2,4-dinitrophenylhydrazine is prepared by the following route:

Chemicals required (other than common laboratory reagents):

Benzene
Bromine
Tin tack
Anhydrous sodium sulphate
Ethanol
Hydrazine hydrate (64 per cent)
Ethyl acetate

Preparation of bromobenzene

Fit a 50 cm³ flask with a reflux condenser and add to the flask 20 cm³ (0.226 mole) of benzene, 6 cm³ (0.11 mole) of bromine (*handle with great care* – use protective gloves), and one small iron tin-tack. Warm the flask with a beaker of water at 50–55 °C. Since hydrogen bromide will be evolved, either the reaction must be carried out in a fume cupboard or some provision must be made to absorb the fumes. In about 15 minutes, after the spontaneous reaction has started to subside, remove the water bath and heat the reaction mixture to boiling. Within 10 minutes the bromine vapours above the liquid in the flask will have disappeared. At this time cool the mixture to room temperature, add 25 cm³ of ether, and extract the solution with two 5 cm³ portions of 2M sodium hydroxide and one 10 cm³ portion of water. Retain the ether extract.

Dry the ether solution over anhydrous sodium sulphate for a few minutes, filter it, and distil it*, collecting the fraction boiling between 140 and 160 °C. The yield should be about 75 per cent. Gas chromatographic analysis shows that the product contains about 4 per cent benzene.

*   Take the usual precautions against fire during the removal of the ether.

Preparation of 2,4-dinitrobromobenzene

Heat a mixture of 36 cm^3 concentrated sulphuric acid and 12 cm^3 concentrated nitric acid in a 100 cm^3 conical flask to 85–90 °C. Add to this 5 cm^3 (0.05 mole) of bromobenzene in three or four portions during a minute, and swirl the mixture well after each addition. The temperature will rise to 130–135 °C. Allow the mixture to stand with occasional swirling for 5 minutes, then cool it nearly to room temperature and pour it over about 150 g of ice. Stir the resulting mixture until the product has solidified, crush the lumps, and collect the crude product by suction filtration.

Recrystallize the product by dissolving it in 30 cm^3 of hot 95 per cent ethanol and allowing the solution to cool. The product usually separates out as an oil, but vigorous swirling of the mixture when the oil appears will promote crystallization. After crystals have started to form, cool the solution in an ice bath. Crystallization should be complete in about five minutes. Collect the product by suction filtration, and wash the pale yellow crystals with a small amount of cold ethanol. The yield should be about 70 per cent of crystals, melting point 69–71 °C.

Preparation of 2,4-dinitrophenylhydrazine

Prepare a solution of 2 g (0.08 mole) of 2,4-dinitrobromobenzene in 30 cm^3 of 95 per cent ethanol, and heat it almost to boiling. Add to this a solution of 2 cm^3 of hydrazine hydrate (0.04 mole of hydrazine) in 10 cm^3 of 95 per cent ethanol (*caution*: use protective gloves). The light orange solution, which rapidly turns a deep red-purple, is allowed to cool undisturbed for 15–20 minutes. Collect the resulting crystals by suction filtration, and wash them with a little 95 per cent ethanol. The yield should be about 90 per cent of crystals having the melting point 200–202 °C. The product sometimes crystallizes as red-purple prisms alone, and sometimes as red-purple prisms and orange plates. Recrystallization of either form from boiling ethyl acetate (50 cm^3 per gramme) affords a product in the form of orange plates (70 per cent recovery).

Answers to problems in the *Students' Book*

(A suggested mark allocation is given in brackets after each answer.)

1 A CH$_3$CHCH$_3$ (1)
          |
          Br
2-bromopropane   (1)

B CH$_3$CHCH$_3$ (1)
          |
          OH
propan-2-ol   (1)

C CH$_3$CCH$_3$ (1)
          ||
          O
acetone   (1)

Total (6)

2 A CH$_3$CH$_2$CH$_2$CH$_2$OH   (1)
butan-1-ol   (1)
Isomer   CH$_3$CHCH$_2$OH   (1)
              |
              CH$_3$
2-methylpropan-1-ol   (1)

B CH$_3$CH$_2$CH$_2$CHO   (1)
butanal   (1)

C CH$_3$CH$_2$CH$=$CH$_2$ (1)
but-1-ene   (1)

D CH$_3$CH$_2$CHCH$_3$ (2)
              |
              I
2-iodobutane   (1)

CH$_3$CH$_2$CCH$_3$ (2)
              ||
              O
butanone   (1)

Total (14)

3 Alternative answers are possible for A, B, and C in (*i*).

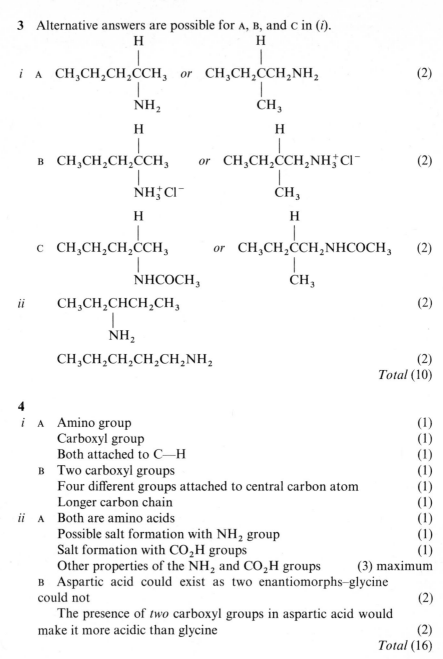

i A $\underset{\underset{\displaystyle NH_2}{|}}{CH_3CH_2CH_2\overset{\overset{\displaystyle H}{|}}{C}CH_3}$ *or* $\underset{\underset{\displaystyle CH_3}{|}}{CH_3CH_2\overset{\overset{\displaystyle H}{|}}{C}CH_2NH_2}$ (2)

B $\underset{\underset{\displaystyle NH_3^+Cl^-}{|}}{CH_3CH_2CH_2\overset{\overset{\displaystyle H}{|}}{C}CH_3}$ *or* $\underset{\underset{\displaystyle CH_3}{|}}{CH_3CH_2\overset{\overset{\displaystyle H}{|}}{C}CH_2NH_3^+Cl^-}$ (2)

C $\underset{\underset{\displaystyle NHCOCH_3}{|}}{CH_3CH_2CH_2\overset{\overset{\displaystyle H}{|}}{C}CH_3}$ *or* $\underset{\underset{\displaystyle CH_3}{|}}{CH_3CH_2\overset{\overset{\displaystyle H}{|}}{C}CH_2NHCOCH_3}$ (2)

ii $\underset{\underset{\displaystyle NH_2}{|}}{CH_3CH_2CHCH_2CH_3}$ (2)

$CH_3CH_2CH_2CH_2CH_2NH_2$ (2)

Total (10)

4

i A Amino group (1)
Carboxyl group (1)
Both attached to C—H (1)
B Two carboxyl groups (1)
Four different groups attached to central carbon atom (1)
Longer carbon chain (1)
ii A Both are amino acids (1)
Possible salt formation with NH_2 group (1)
Salt formation with CO_2H groups (1)
Other properties of the NH_2 and CO_2H groups (3) maximum
B Aspartic acid could exist as two enantiomorphs–glycine could not (2)
The presence of *two* carboxyl groups in aspartic acid would make it more acidic than glycine (2)

Total (16)

5 (5) for an acceptable method
The most likely route would be:

$$\underset{\substack{\text{CH}-\text{CH}\\ \parallel \qquad \parallel\\ \text{CH} \qquad \text{C}-\text{CH}_2\text{OH}\\ \diagdown \quad \diagup\\ \text{O}}}{} \xrightarrow{\text{H}_2} \underset{\substack{\text{CH}_2-\text{CH}_2\\ \diagup \qquad \diagdown\\ \text{CH}_2 \qquad \text{CH}-\text{CH}_2\text{OH}\\ \diagdown \quad \diagup\\ \text{O}}}{} \xrightarrow{\text{SOCl}_2}$$

$$\underset{\substack{\text{CH}_2-\text{CH}_2\\ \diagup \qquad \diagdown\\ \text{CH}_2 \qquad \text{CH}-\text{CH}_2\text{Cl}\\ \diagdown \quad \diagup\\ \text{O}}}{} \xrightarrow{\text{Na}} \text{final product}$$

6 3 marks for each part

7 Parts (*i*) to (*iv*):
4 marks each for an acceptable method
Look for the minimum number of separate steps for each change

Topic 14
Reaction rates

Objectives

1   To study the factors which influence the rate of reaction.

2   To show that the course of a reaction is often complex, and that it may proceed by several stages.

3   To introduce a simple treatment of theories of reaction rates.

Content

14.1   Introduction.

14.2   The kinetics of the reaction between iodine and acetone in aqueous solution.

14.3   The hydrolysis of bromoalkanes.

14.4   Theories of reaction kinetics.

14.5   An investigation of the kinetics of another reaction.

Timing

Three weeks.

14.1 **Introduction**

Objectives
1 To see how the kinetics of a reaction can be followed.
2 To decide what is meant by 'rate of a reaction'.
3 To see what factors affect this rate.

Timing
One double period will be sufficient if students are already familiar with the factors that influence the rate of a reaction.

Suggested treatment
It is suggested that students should read the introduction to this topic for homework and that it should be discussed in class. It covers the reasons for studying kinetics; some of the methods of following a reaction to determine its rate; and a possible definition of 'the rate of a reaction'. It is best to define the rate of a reaction in such a way that the definition is independent of factors such as the volume of the reaction system, the number of phases present, or which reactant or product is being followed. But at this level a more simple definition such as the rate of disappearance of one of the reactants or the rate of appearance of one of the products is preferable. As this definition is used it should be referred to by some phrase such as 'the rate of the reaction expressed as the rate of increase of concentration of B', for, as is pointed out in the *Students' Book*, it is most important to state what B is.

(For a fuller discussion, see *Physico-chemical quantities and units* by M. L. McGlashan, Monograph for Teachers No. 15 (1968), published by the Royal Institute of Chemistry.)

The discussion of what factors affect the rate of the reaction will depend on the previous experience of the students. If they are already familiar with them it may be sufficient to remind them of experiments such as those with sodium thiosulphate solution and acid which are described on page 336 of Nuffield Chemistry: *The Sample Scheme Stages I and II, The Basic Course*. These experiments show how the concentrations of reactants and temperature affect the rate of the reaction

$$S_2O_3^{2-}(aq) + 2H^+(aq) \rightarrow S(s) + SO_2(aq) + H_2O(l)$$

If students have no previous experience of reaction rates, it may be more appropriate for them to do these experiments themselves.

Suggestions for homework
Reading the introduction to this topic in the *Students' Book*.

Summary
From this section students should appreciate: why it is useful to study kinetics; some of the ways in which the rate of a reaction can be followed; what is meant by the term 'rate of reaction'; and that increase of concentration of reactants and increase of temperature increase the rate of a reaction.

14.2   The kinetics of the reaction between iodine and acetone in aqueous solution

Objectives
To investigate the factors that affect the rate of the reaction between iodine and acetone in solution; to establish the rate expression for the reaction; and to show that the stoichiometric equation tells us little of what the rate may depend upon.

Timing
Approximately one week.

Suggested treatment
The objectives mentioned above can be achieved by a study of a number of different reactions in several different ways. One study is described here, and an alternative indicated at the end of the section.

How different concentrations of acetone, iodine, and acid affect the rate of the reaction

$$CH_3COCH_3(aq) + I_2(aq) \rightarrow CH_3COCH_2I(aq) + H^+(aq) + I^-(aq)$$

can be found by determining the initial rate of the reaction under different conditions. The rate can be found in terms of rate of decrease of concentration of iodine, this concentration being determined by a colorimeter at suitable intervals of time; or alternatively, by means of a titration technique. Details of the colorimetric method are given in the *Students' Book*.

Experiment 14.2
An investigation of the rate of the reaction between iodine and acetone in aqueous solution

Each pair of students will need:

1 stopclock, or sight of a large clock with seconds hand
1 test-tube
1 colorimeter, with test-tubes

access to communal burettes containing:

0.01 M iodine solution (2.54 g iodine and 8 g potassium iodide made up to 1 dm^3 with water)
2M acetone (142 cm^3 acetone made up to 1 dm^3 with water)
2M hydrochloric acid
Water

Notes. These concentrations, as all in this topic, are for a specific colorimeter. Thus the iodine concentration may require modification for the colorimeter used.* The acetone and/or acid concentrations may be modified so that the reaction proceeds more quickly or slowly if desired.

It is worth reminding students that iodoacetone may make their eyes water, although, in these concentrations, it is unlikely to prove hazardous.

Procedure

Students carry out an experiment showing that when iodine and acetone solutions are mixed, little happens; but on addition of hydrochloric acid, the solution loses its colour. As iodine is known not to react with hydrochloric acid, it must have reacted with the acetone, and the acid must have affected the rate of this reaction in some way.

How the concentrations of acetone, iodine, and hydrogen ions affect the rate of the reaction is then investigated. The initial rate is found for varying concentrations of acetone, iodine, and hydrogen ions. *It is important to note* that only the initial rate is required in this experiment; the form of the rest of the curve is not important here. Students will probably find, with the limited number of points that they plot, that the curve is a straight line. The different mixtures, details of which are given in the *Students' Book*, should be investigated by different pairs of students so that the class investigates them all and each one is investigated at least twice so that some idea of the errors involved can be obtained.

*For details of the colorimeter, see Appendix 2.

It will be found, within experimental error, that the rate is proportional to the concentration of acetone:

rate $\propto$ [acetone]

or

$$\text{rate} = k\,[\text{acetone}]$$

$$\left[\text{or} \quad -\frac{d[I_2]}{dt} = k\,[\text{acetone}]\right]$$

and the numerical value of the rate constant can be determined. The rate is in terms of decrease in concentration of iodine in unit time and therefore in the above equation, the units on the lefthand side are

$\text{mol dm}^{-3}\,\text{s}^{-1}$

and those of the righthand side are

$\text{mol dm}^{-3} \times$ (units of k).

Thus the units of k must be s^{-1}.

It will also be found that the rate is not affected by the iodine concentration, and that it is proportional to the hydrogen ion concentration:

$$\text{rate} = k_1[\text{H}^+]$$

An overall expression for the rate is therefore:

$$\text{rate} = k_2\,[\text{acetone}]\,[\text{H}^+]$$

The units of the lefthand side of this equation are:

$\text{mol dm}^{-3}\,\text{s}^{-1}$

and those of the righthand side are

$\text{mol dm}^{-3} \times \text{mol dm}^{-3} \times$ (units of k_2).

Thus the units of k_2 must be $\text{dm}^3\,\text{mol}^{-1}\,\text{s}^{-1}$.

Several important points should be made at this stage.

1 The above equation is *the rate expression* for the reaction: the mathematical equation involving the concentrations of the reagents and a constant, which is found *experimentally*, to express the rate.

2 The stoichiometric equation tells us nothing about the rate expression, what concentrations appear in it, and to what power they are raised. As can be seen in this example, the concentration of iodine has no effect upon the rate and does not appear in the rate expression; and the hydrogen ion concentration does not appear on the lefthand side of the stoichiometric equation and yet does appear in the rate expression.

3   The idea of *the order of the reaction* should be discussed. With respect to a particular reactant, it is with the power to which its concentration is raised in the rate expression. This reaction is first order with respect to acetone and first order with respect to hydrogen ions (and zero order with respect to iodine) and so the *overall order* can be said to be second order. It should be emphasized that the order of a reaction is an experimentally determined fact.

4   The *rate constant* is the constant in the rate expression; it is dependent on temperature, and for a first order reaction has units of s^{-1}; for a second order reaction the units are $dm^3\,mol^{-1}\,s^{-1}$.

5   The rate expression may not be as simple as this one. Most reactions occur in several steps and the power to which concentrations are raised need not be a whole number. An example of a fractional power is found in the rate expression for the formation of hydrogen bromide from hydrogen and bromine:

$$\text{rate} = \frac{k[H_2][Br_2]^{\frac{3}{2}}}{[Br_2] + k'[HBr]}$$

Problem 1 in the *Students' Book* gives some rate/concentration results for the decompôsition of dinitrogen pentoxide in tetrachloromethane solution. It is found that the rate is proportional to the concentration of dinitrogen pentoxide, and a value for the rate constant can be obtained.

   Answers to problem 1
1   Rate = constant $\times [N_2O_5]$
2   First order
3   $1.07 \times 10^{-5}\,s^{-1}$.

Problem 2 in the *Students' Book* gives data to enable students to plot an extent of reaction/time graph for the reaction:

$$2H^+(aq) + H_2O_2(aq) + 2I^-(aq) \rightarrow I_2(aq) + 2H_2O(l)$$

It should be pointed out to the students that the concentrations of the reactants fall as the reaction proceeds, so the rate of the reaction decreases and the graph flattens out. From gradients taken from the graph (these can be taken accurately so long as the graph is smooth) students can plot a rate/concentration of hydrogen peroxide graph which will indicate that the reaction is first order. A further interesting fact about first order reactions can be obtained from the concentration/time graph; the times taken for the concentration to fall to one half of its initial value, from one half to one quarter, from one quarter to one eighth, and from one eighth to one sixteenth of it, will be found to be the same – the half-life of the reaction. Thus *for a first order reaction* the half-life is constant and independent of the concentration at the beginning of each half-life.

Answers to problem 2

Reaction is first order

$$t_{\frac{1}{2}} = 3.9 \times 10^3 \text{ s}$$

Problem 3 gives data for the decay of Iodine-138 and from it students can show that the half-life of this decay is also constant, so that radioactive decays follow first order kinetics.

Answers to problem 3

$$\left. \begin{array}{l} 1 \\ 2 \\ 3 \end{array} \right\} t_{\frac{1}{2}} = 1.44 \times 10^3 \text{ s}$$

4   First order

If teachers wish they may introduce the integrated form of the first order rate equation in problem 2 to show that the order of the reaction and the rate constant can be obtained more simply in this way. It should not be introduced before, as one has to assume that the rate is proportional to concentration to obtain it, and students should find this relationship for themselves in experiment 14.2. Unless the class will understand the integration properly, it is best not to introduce it to them. For this reason, integrations of both the first and second order rate equation (and the connection between half-life and rate constant) are given in an Appendix in the *Students' Book*, and not in the main body of the text.

Problem 4 gives data about a second order reaction which enables students to find that half-lives in this case are not equal but depend on the concentrations at the beginning of the half-life.

Answers to problem 4

The half-lives are

1 $1.98 \times 10^3 \text{ s}$
2 $4.68 \times 10^3 \text{ s}$
3 $7.92 \times 10^3 \text{ s}$

Alternative method for experiment 14.2

If colorimeters are not available, the iodine concentrations can be determined by titration. Mixtures are made up according to the table in the *Students' Book*, but using 15 times the volumes giving a total of 150 cm^3 of mixture. 25 cm^3 portions are removed in such a way as to be discharged into about 30 cm^3 of saturated sodium bicarbonate solution at each minute. The mixture is swirled till effervescence ceases and is then titrated with 0.005M sodium thiosulphate solution using 0.2 per cent starch solution as indicator.

Alternative reaction for experiment 14.2

The reaction between hydrogen peroxide and iodide ions in acid solution

$$H_2O_2(aq) + 2H^+(aq) + 2I^-(aq) \rightarrow 2H_2O(l) + I_2(aq)$$

can be studied instead of the iodination of acetone. The rate of formation of iodine can be followed in a colorimeter, and the concentrations of hydrogen peroxide, iodide ions, and hydrogen ions can be investigated. Initial rates are found for the mixtures in table 14.2.

	Volume 0.1M KI /cm^3	Volume 0.5M H$_2$SO$_4$ /cm^3	Volume 0.1M H$_2$O$_2$ /cm^3	Volume water /cm^3	Total volume /cm^3
a	3	3	3	3	12
b	2	3	3	4	12
c	1	3	3	5	12
d	3	2	3	4	12
e	3	1	3	5	12
f	3	3	2	4	12
g	3	3	1	5	12

Table 14.2

If colorimeters are not available, 1 cm^3 of 0.01M sodium thiosulphate and 1 cm^3 0.2 per cent starch solution are added to each mixture, and the time for a fixed amount of iodine to be produced (that which reacts with 1 cm^3 0.01M $S_2O_3^{2-}$(aq)) is found by noting the time for the mixture to turn blue. This reflects the rate of the reaction under different conditions. The points indicated above can be made on the basis of the results of this experiment:

> rate $\propto$ [hydrogen peroxide]^1
> rate $\propto$ [iodide ions]^1 (In problem 2 [I$^-$(aq)] is constant.)

and the rate is, for the most part, independent of the acid concentration:

> rate $\propto$ [hydrogen ions]^0

(In fact two pathways are available for this reaction and

> rate $= k[H_2O_2][I^-] + k'[H^+][H_2O_2][I^-]$

but $k' \ll k$.)

Suggestions for homework

Problems 1 to 5.

Summary
Students should understand what is meant by:

> rate expression
> rate constant
> reaction order
> half-life

and should realize that the stoichiometric equation tells us nothing of what concentrations will be involved in the rate expression.

14.3   **The hydrolysis of bromoalkanes**

Objectives
To find out that the hydrolysis of the three bromoalkanes of formula C_4H_9Br occurs at very different rates, and to suggest an explanation for this in terms of different mechanisms.

Timing
Two double periods.

Suggested treatment
Students should be reminded of the experimental work they did on the hydrolysis of halogenoalkanes in experiment 9.5a, and then should carry out similar experiments with the three bromoalkanes.

Experiment 14.3
A comparison of the rates of hydrolysis of some bromoalkanes
Each student or pair of students will need:

> 3 hard glass test-tubes and rack
> Teat pipette

access to:

> Approximately 2M sodium hydroxide solution
> Approximately 0.1M silver nitrate solution
> Ethanol (industrial methylated spirits)
> 1-bromobutane
> 2-bromobutane
> 2-bromo-2-methylpropane

Procedure
Students mix 1 cm³ of ethanol, 2 drops of a bromoalkane, and 1 cm³ of silver nitrate solution in each of the test-tubes, using a different bromoalkane in each tube. They then observe the mixtures for at least five minutes. It will probably be necessary to suggest to students that the type of observations they should make are 'no observable reaction after two minutes' or 'faint precipitate after six minutes'.

Typical results at 20 °C for this experiment are as follows:

1-bromobutane – Very slight opalescence after about three minutes, gradually intensifying, but not becoming opaque even after 15 minutes.

2-bromobutane – Slight opalescence after only 15 seconds, becoming moderately opaque after one minute. Some coagulation and precipitation after four minutes.

2-bromo-2-methylpropane – Immediate opaque yellow suspension on adding silver nitrate solution, with some yellow precipitate appearing in two minutes.

Discussion of these results could be along the following lines; to aid the discussion the structural formulae given below are also included in the *Students' Book*. The Nuffield film loop 'Hydrolysis of bromoalkanes' should be shown at a suitable point in the discussion.

There are two possible ways in which this hydrolysis could occur.

1   The bromobutane might first ionize:

$$
\begin{array}{ccc}
CH_3 & & CH_3 \\
| & & | \\
CH_2 & & CH_2 \\
| & & | \\
CH_2 & \longrightarrow & CH_2 \\
| & & | \\
\underset{H}{\overset{}{C}}\,|\,Br & & \underset{H}{\overset{}{C^+}}\,|\,+Br^- \\
H & & H
\end{array}
$$

and then the positive ion so formed react with an OH^- ion to give the alcohol:

$$
\begin{array}{ccc}
CH_3 & & CH_3 \\
| & & | \\
CH_2 & & CH_2 \\
| & \longrightarrow & | \\
CH_2 & & CH_2 \\
| & & | \\
\underset{H}{\overset{}{C^+}}\,+OH^- & & \underset{H}{\overset{}{C}}\,|\,OH \\
H & & H
\end{array}
$$

2   Alternatively, attack by the hydroxide ion and ejection of a bromide ion might happen in such a way that for an instant of time both the incoming hydroxide and outgoing bromide ions would be equally associated with the hydrocarbon group in a *transition state*:

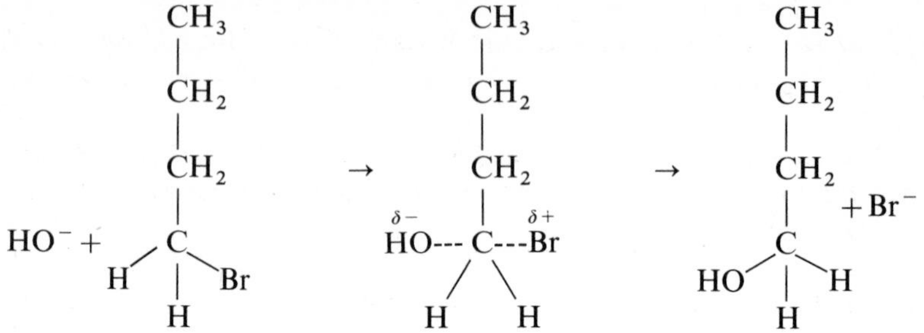

Is it possible to find out which route a particular bromoalkane hydrolysis takes? The rates of hydrolysis of 2-bromo-2-methylpropane and of the other two compounds are so different that it may be that these reactions take different routes.

If the route taken is (2) then the reaction starts when an OH^- attacks the bromobutane, and therefore we would expect that both the concentration of OH^- and that of bromobutane would affect the hydrolysis. In other words, it might be expected that the rate of reaction is proportional to both these concentrations.

Expressed in symbols

$$\text{rate} \propto [RBr] \quad \text{and} \quad \text{rate} \propto [OH^-]$$

that is, rate $\propto [RBr][OH^-]$.

If the route taken is (1) we might expect the second reaction to be fast as most reactions between ions are fast:

$$Pb^{2+}(aq) + 2Cl^-(aq) \rightarrow PbCl_2(s)$$
$$H^+(aq) + OH^-(aq) \rightarrow H_2O(l)$$
$$R^+(aq) + OH^-(aq) \rightarrow ROH(aq)$$

The first reaction however will probably be slow as it involves breaking a bond. Therefore this reaction is the 'bottleneck' in the process or the *rate determining step*, and as it only involves RBr and not OH^-, we would expect that for this route, the rate would only be affected by the concentration of RBr and not by that of OH^-, that is,

$$\text{rate} \propto [RBr]$$

Experiments have been done which show that for 2-bromo-2-methylpropane

rate $\propto$ [RBr]

showing that the two-stage route, (1), is followed; and for 1-bromobutane

rate $\propto$ [RBr][OH$^-$]

showing that the one-stage route, (2), is followed. For 2-bromobutane, *both* routes are followed as the rate is proportional to [RBr] but to a fractional power of [OH$^-$].

It could then be suggested that the experiments with 1-bromobutane and 2-bromobutane be repeated with sodium hydroxide solution instead of silver nitrate solution, acidifying and adding silver nitrate solution after a suitable time. It will be found that the hydrolysis is quite rapid in the presence of alkali.

Is the one-stage route possible for 2-bromo-2-methylpropane? Students can be asked to look at a model of this molecule and find out. It is obvious that sterically it is impossible for the OH$^-$ to attack before the bromine atom has left. The second half of the Nuffield film loop 'The hydrolysis of bromoalkanes' demonstrates this point.

When this experiment is over, students can be set problem 5, given at the end of the section. This uses the idea of a rate-determining step. The data provide information which should enable the students to deduce that the rate-determining step must involve

$$2H^+(aq) + BrO_3^-(aq) + Br^-(aq)$$

Answers to Problem 5

A and B indicate that when the [H$^+$] is doubled, the rate quadruples showing

rate $\propto$ [H$^+$]^2

B and C indicate that when [BrO$_3^-$] is doubled, the rate is doubled showing

rate $\propto$ [BrO$_3^-$]

D and B indicate that when the [Br$^-$] is doubled the rate is doubled showing that

rate $\propto$ [Br$^-$]

This shows (as might be expected from the stoichiometric equation) that the reaction occurs in steps, and the rate determining step(s) involve all three reagents.

At this stage students may want to look for a mechanism in the iodine–acetone and iodide–hydrogen peroxide reactions.

The iodine-acetone reaction

The rate suggests that the 'bottleneck' in the reaction is a step involving one acetone molecule and one hydrogen ion only. The suggested first slow step in the reaction is:

$$\underset{\substack{\\ \\ CH_3-\overset{\displaystyle O}{\overset{\displaystyle \|}{C}}-CH_3}}{H^+} \rightarrow \underset{\substack{\\ \\ H}}{\overset{\substack{H\\ |}}{H-\overset{\displaystyle O}{\overset{\displaystyle |}{C}}-\underset{+}{C}-CH_3}}$$

followed by a fast reaction in which a hydrogen ion is lost:

$$\underset{\substack{\\ H}}{\overset{\substack{H\\ |\\ H\ \ O\\ |\ \ |}}{H-C-\underset{+}{C}-CH_3}} \rightarrow \underset{\substack{\\ H}}{\overset{\substack{H\\ |\\ H^+\ \ O\\ |}}{H-C=C-CH_3}}$$

This gives the enol form of the ketone. In some ketones this enol form is present to a great extent, but in acetone, the keto form predominates and the enol form is present only to an extent of 0.00025 per cent. This reacts like any carbon-carbon double bond with a halogen:

$$\underset{\substack{\\ I_2}}{\overset{\substack{OH\\ |}}{CH_2=C-CH_3}} \rightarrow \underset{\substack{\\ I\ \ \ I}}{\overset{\substack{OH\\ |}}{CH_2-C-CH_3}}$$

and the compound so formed breaks up to give the keto form back again and hydrogen and iodide ions:

$$\underset{\substack{\\ I\ \ \ I}}{\overset{\substack{OH\\ |}}{CH_2-C-CH_3}} \rightarrow \underset{\substack{\\ I^-}}{\overset{\substack{H^+\\ \\ O\\ \|}}{CH_2I-C-CH_3}}$$

The iodide-hydrogen peroxide reaction

The rate expression suggests that the 'bottleneck' in this reaction is a step involving one iodide ion and one hydrogen peroxide molecule only. The suggested first slow step in the reaction is:

$$H_2O_2(aq) + I^-(aq) \rightarrow H_2O(l) + OI^-(aq)$$

followed by two fast reactions:

$$H^+(aq) + OI^-(aq) \rightarrow HOI(aq)$$
$$HOI(aq) + H^+(aq) + I^-(aq) \rightarrow I_2(aq) + H_2O(l)$$

It should be made clear that the whole of these mechanisms cannot be deduced from these single pieces of kinetic evidence, only that the *rate-determining steps* involve acetone molecules and hydrogen ions, and hydrogen peroxide molecules and iodide ions respectively; but together with other evidence, these have led to the suggestion of the above mechanisms. Moreover, any proposed mechanism is speculative to some extent and is in fact a model built up from such evidence.

After section 14.5 has been studied, it would be profitable to ask students how they think the rate of iodination of acetone will vary with time, if one starts with acetone and iodine only. As hydrogen ions are produced, this reaction is also autocatalytic.

Supporting material

Nuffield film loop 'The hydrolysis of bromoalkanes'.

Summary

Students should realize that it may be possible to obtain some idea of the mechanism of a reaction from a study of the rate of the reaction.

14.4   Theories of r₁ action kinetics

Objectives

To look at the collision theory in a qualitative manner, to see its limitation, and to note briefly the transition state theory of kinetics.

Timing

Three periods should be sufficient.

Suggested treatment

It is important that this section should not be treated in a rigorous mathematical way (unless the teacher thinks that the class would prefer this), but should aim to show the way in which these types of calculation can be made and the sort of results that can be obtained from them. The Nuffield film loop 'Rate of reaction' can be shown at an appropriate point in this treatment.

Collision theory

Discussion of the collision theory of reactions could start by supposing that the rate of reaction is similar to the rate of collision of the particles that are reacting. It is difficult to quote results for an actual reaction, as it is virtually impossible to find a reaction which proceeds solely by a single step:

$A(g) + B(g) \rightarrow$ products

but many gas reactions which are predominantly second order have a rate constant of about the same magnitude.

Some such reactions are:

$2HI(g) \rightarrow H_2(g) + I_2(g)$
$H_2(g) + I_2(g) \rightarrow 2HI(g)$
$2NO_2(g) \rightarrow 2NO(g) + O_2(g)$
$2NOCl(g) \rightarrow 2NO(g) + Cl_2(g)$
$NO_2(g) + O_3(g) \rightarrow NO_3(g) + O_2(g)$

$$
\begin{array}{ccc}
CH_2 & CH_2 & CH_2 \\
\| & \| & / \quad \backslash \\
CH & CH & CH \quad\quad CH_2 \\
| \quad + \quad | & \rightarrow & \| \quad\quad | \\
CH & CHO & CH \quad\quad CH \\
\| & & \backslash \quad / \quad \backslash \\
CH_2 & & CH_2 \quad CHO
\end{array}
$$

We will therefore consider a hypothetical gas reaction of the type

$2A(g) \rightarrow$ products,

where A has molecular weight 125, taking place at 773 K with an initial value for $[A(g)]$ of 10^{-2} mol dm^{-3}, for which $k = 10^{-2}$ dm^3 mol^{-1} s^{-1}.

Two pieces of information are needed:
    the total number of collisions per second in a given volume
    the rate of reaction in terms of the number of effective collisions per second in the same volume.
The common volume will be taken as one cubic metre.

The total number of collisions can be found by considering a cylinder of twice the average diameter of the molecules of A, and v metres long. A molecule travelling at v m s^{-1} will collide with all the particles whose centres are within this cylinder during one second.

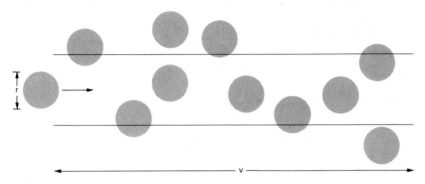

Figure 14.4a

The average radius of the molecules and their average velocity can be found. The total number of molecules within the cylinder, which equals the number of molecules with which the chosen particle will collide per second, can be calculated from the original concentration of the gas. The result can be multiplied up to account for all the particles, a process in which each collision must be considered twice, so the answer must be divided by two to allow for this, and a value for the total number of collisions found.

For molecules of radius 0.4 nm, with molecular weight 125 and average velocity 3.6×10^2 m s^{-1} at 773 K, *the total number of collisions per cubic metre per second* $= 3.2 \times 10^{33}$.

Semi-rigorous derivation, if required
A collision between two identical molecules will occur if their centres approach within a distance equal to their average diameter. This distance can be found from gas viscosity measurements. A molecule of diameter 0.4 nm, travelling at a speed of v m s^{-1} will cover v metres in one second and during this second will collide with every molecule whose centre is within a cylinder of length v cm and *radius* 0.4 nm.

The volume of this cylinder $(\pi r^2 l)$ is $3.14 \, (0.4 \times 10^{-9})^2 \, v \, m^3 = 5 \times 10^{-19} \, v \, m^3$.

When $[A(g)] = 10^{-2}$ mol dm^{-3} the total number of molecules available *per cubic metre* for collision with the given molecule is

$$10^{-2} L \times 10^3 \, (L, \text{Avogadro constant}, = 6 \times 10^{23} \text{ mol}^{-1}; 1 \text{ m}^3 = 10^3 \text{ dm}^3)$$
$$= 10^{-2} \times 6 \times 10^{23} \times 10^3 = 6 \times 10^{24}$$

Thus the number of molecules within the cylinder

$$= 6 \times 10^{24} \times 5 \times 10^{-19} \, v = 3 \times 10^6 \, v$$

and the given molecule will collide with all these in one second. The total number of molecules which can collide in this way is 6×10^{24} per cubic metre, hence the total number of collisions per cubic metre per second

$$= \tfrac{1}{2}(3 \times 10^6 \, v)(6 \times 10^{24}) = 9 \times 10^{30} \, v$$

(The factor $\tfrac{1}{2}$ corrects for counting each collision twice in this process; we have counted A_1 colliding with A_2 and also A_2 colliding with A_1, which is the same collision.)

We now require the average velocity of the molecules, from kinetic theory of gases; this is given by

$$v = \left(\frac{8 \times 10^3 \, RT}{\pi M}\right)^{\frac{1}{2}}$$

where M = molecular weight
and $R = 8.3$ J mol^{-1} K^{-1} (the gas constant)
at 773 K with $M = 125$

$$v = \left(\frac{8 \times 10^3 \times 8.3 \times 773}{3.14 \times 125}\right)^{\frac{1}{2}} = 3.6 \times 10^2 \text{ m s}^{-1}$$

The total number of collisions per cubic metre per second

$$= 9 \times 10^{30} \times 3.6 \times 10^2 = 3.2 \times 10^{33}$$

(*Note.* In this treatment some assumptions have been made about the velocity of the given molecule, for example that it does not deviate from a straight line path throughout the v metres of its travel. A more rigorous treatment gives a value of 3.6×10^{33} collisions m^{-3} s^{-1}.)

The rate of the reaction in terms of the number of effective collisions (those which lead to reaction) can be calculated quite easily. For $[A(g)] = 10^{-2}$ mol dm^{-3} and $k = 10^{-2}$ dm^3 mol^{-1} s^{-1} it is *number of effective collisions per cubic metre per second* $= 3 \times 10^{20}$.

Rigorous derivation, if required

rate of reaction $= k \, [A(g)]^2$
$$= 10^{-2} \times (10^{-2})^2 = 10^{-6} \text{ mol dm}^{-3} \text{ s}^{-1}$$

number of molecules decomposing per second per cubic decimetre
$$= 10^{-6} \, L = 10^{-6} \times 6 \times 10^{23}$$

but 1 m^3 = 10^3 dm^3

∴ number of molecules decomposing per second *per cubic metre*

$$= 10^{-6} \times 6 \times 10^{23} \times 10^3$$
$$= 6 \times 10^{20}$$

Two molecules are required to effect one collision.
∴ number of effective collisions per cubic metre per second

$$= \tfrac{1}{2} \times 6 \times 10^{20} = 3 \times 10^{20}$$

Thus only a minute percentage $\left(\dfrac{3 \times 10^{20} \times 100}{3.2 \times 10^{33}} \approx 10^{-11} \text{ per cent}\right)$ of the colli-

sions result in reaction. Let us consider the reaction more closely. When collision occurs, what happens next? Whether the reaction is a decomposition or any other type, normally the first occurrence is the breaking of bonds. Other bonds may be made later and the overall energy may be exothermic or endothermic, but before other bonds can be made, some must be broken. Obviously this initial breaking of bonds requires energy and therefore only those collisions, after which the particles have sufficient energy for these bonds to be broken, will result in reaction. The fact that certain initial minimum energy is required is demonstrated by gunpowder, coal, a match and many other examples of exothermic reactions which nevertheless require an initial input of energy.

If this is right, it becomes important to find what fraction of the collisions have this certain minimum energy. First, what value should be used for the minimum energy? Before the bond is completely broken, a new bond will have started to form so the energy required would not be quite as large as the average of bond energy terms. These vary from about 150 kJ mol^{-1} for the weak I—I bond to about 550 kJ mol^{-1}, for strong bonds such as H—F with other single bond values intermediate (O—H, 460 kJ mol^{-1}; C—C, 350 kJ mol^{-1}). With the average single bond value at around 350 kJ mol^{-1}, a reasonable guess might be 200 kJ mol^{-1}.

How can the fraction of the collisions that have an energy greater than 200 kJ mol^{-1} be calculated? The energy of the particles is mainly a function of their velocity and the distribution of velocities can be experimentally determined by an apparatus such as that used by Zartman in 1931. Although the velocities of particles are continually changing due to collision, the fraction of particles with a certain velocity remains constant. Zartman's technique can best be explained by means of a film loop: Longman's chemistry film loop no. 8 'Experiments to show Maxwell-Boltzmann distribution of velocities and energies in molecules of gas' is suitable. The graph in figure 14.4b shows how the velocities are distributed.

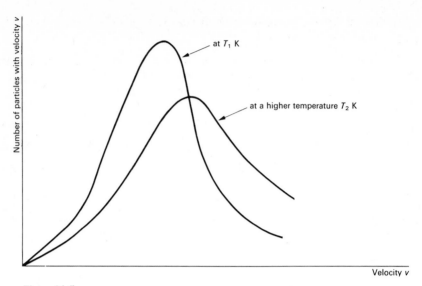

Figure 14.4b

This distribution can also be calculated theoretically and the fraction of collisions with an energy greater than a value E J mol^{-1} is given by:

$$\lg \left(\frac{\text{number of collisions with energy } E}{\text{total number collisions}} \right) = - \frac{E}{2.3 \ RT}$$

$$* \left[or \left(\frac{\text{number of collisions with energy } E}{\text{total number collisions}} \right) = e^{-E/RT} \right]$$

The fraction of the molecules with energy greater than 200 kJ mol^{-1} at 773 K is given by

$$\lg (\text{fraction}) = - \frac{E}{2.3 \ RT} = - \frac{200 \times 10^3}{2.3 \times 8.3 \times 773}$$

$$= -13.6 \ (\text{or } \overline{14}.4)$$

From which the fraction of collisions with total energy greater than 200 kJ mol$^{-1} = 2.5 \times 10^{-14}$.

(This corresponds to one collision in 4×10^{13}.)

---

*In the *Students' Book*, the expression is given in the form not involving e as many students find this type of expression incomprehensible. Teachers may wish to use the neater expression involving e.

Thus the rate at which collisions occur having an energy greater than 200 kJ mol^{-1}

    = total number of collisions per cubic metre per second × fraction with energy > 200 kJ mol^{-1}
    = $3.2 \times 10^{33} \times 2.5 \times 10^{-14}$
    = 8×10^{19} m^{-3} s^{-1}

This is not far off the number of effective collisions calculated above, 3×10^{20} m^{-3} s^{-1}. To make the two results identical a value of 192 kJ mol^{-1} for E is needed, so the original guess was not too bad.

This minimum energy, E, is called the *Energy of activation*. The reaction can be imagined as proceeding in something of the manner shown in figure 14.4c.

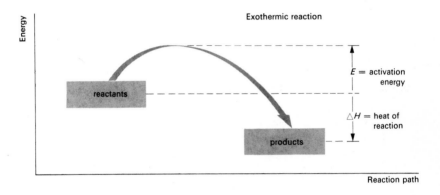

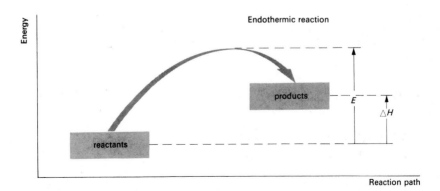

Figure 14.4c

Suppose the rate of the reaction depends on temperature according to the relation

$$\text{rate of reaction} \propto 10^{-E/2.303RT}$$

The rate constant, k, is a measure of the rate of the reaction, independent of concentration, so

$$k \propto 10^{-E/2.303RT} \quad [k \propto e^{-E/RT}]$$

$$\therefore \qquad k = \text{constant} \times 10^{-E/2.303RT}$$

$$\therefore \qquad \lg k = \lg \text{constant} + \lg 10^{-E/2.303RT}$$

since the logarithm of a constant is also constant we have

$$\lg k = c - \frac{E}{2.303R} \times \frac{1}{T}$$

Comparing this with

$$y = c + mx$$

means that a graph of $\lg k$ against $\dfrac{1}{T}$ should, if the theory is correct, be a straight line, and its gradient will be

$$-\frac{E}{2.303R}$$

where E = activation energy in J mol^{-1} and $R = 8.3$ J K^{-1} mol^{-1}. Students can find out if this is true and then obtain a value for the activation energy by a graphical method. To do this values of the velocity constant for given reactions at different temperatures are required. Problem 6 gives such data for the decomposition of hydrogen iodide and the activation energy is found to be 185 kJ mol^{-1}. Problem 7 gives data for the decomposition of benzene diazonium chloride, leading to a value for the activation energy of 115 kJ mol^{-1}. The connection between this lower value and the rate of the reaction should be pointed out.

This simple picture of collision can only be a useful model in certain cases. Difficulties arise immediately we think of a first order reaction and, indeed, very few reactions are second order one-step processes as in our simple model. Moreover this simple model cannot be applied to reactions in solution. Nevertheless the idea of activation energy is a useful one in all reactions since it can be obtained experimentally, and need not be derived by application of a collision theory.

Transition state theory

The collision theory focuses attention on what happens *before* the collision; the transition state theory attacks the problem from another angle and focuses attention on what happens *after* the collision.

A transition state has been met already in the hydrolysis of 1-bromobutane which, it is suggested, proceeds through the transition state:

$$
\begin{array}{c}
\text{H} \\
\text{CH}_3\text{CH}_2\text{CH}_2 \diagdown \diagup \\
\qquad\qquad -\text{C}- \; - \; \text{OH} \\
\text{Br} \; - \; - \; - \; - \; \; | \\
\qquad\qquad\quad \text{H}
\end{array}
$$

Another example is the decomposition of hydrogen iodide. This can be considered to proceed mainly in the following manner:

$$
\begin{array}{c}
\text{H—I} \\
\\
\rightleftharpoons \qquad
\begin{array}{c}
\text{H---I} \\
| \\
| \\
| \\
\text{H---I}
\end{array}
\quad \rightarrow \quad
\begin{array}{c}
\text{H} \\
| \\
\text{H}
\end{array} +
\begin{array}{c}
\text{I} \\
\\
\text{I}
\end{array}
\quad \rightarrow \quad
\begin{array}{c}
\text{H} \\
| \\
\text{H}
\end{array} +
\begin{array}{c}
\text{I} \\
| \\
\text{I}
\end{array} \\
\\
\text{H—I}
\end{array}
$$

The activated complex (designated by the superscript $\ddagger$) is obviously of high energy and therefore can be considered to be at the top of the reaction path diagram given in figure 14.4d.

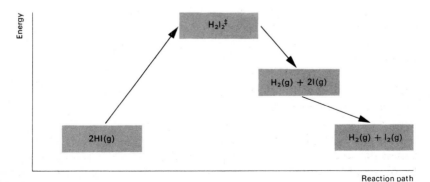

Figure 14.4d

The rate of the reaction can be considered in terms of the rate at which the activated complex, $H_2I_2^{\ddagger}$ decomposes into $H_2(g)$ and $2I(g)$ and then to $I_2(g)$. The calculation is simplified by assuming that the reactant, $HI(g)$, and the activated complex are in equilibrium and that this equilibrium is disturbed as the activated complex decomposes to give products. Calculations of rate constant by this theory have been made and it is more generally applicable than the collision theory.

It is useful to point out the difference between the two routes for the bromoalkane hydrolysis. The two-stage route consists of two definite reactions, as shown in figure 14.4e.

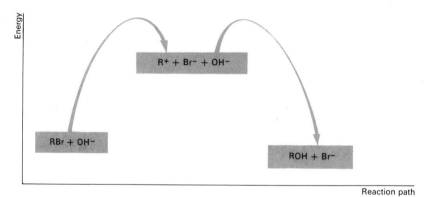

Figure 14.4e

The one-stage route consists of only one reaction, the transition state existing fleetingly on the top of the hump of the curve as in figure 14.4f.

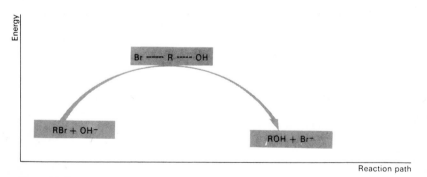

Figure 14.4f

It should be pointed out that having reached the intermediate R^+, or the transition state Br—R—OH, the reaction can go back to RBr as well as on to ROH.

Supporting material
Nuffield film loop 'Rate of reaction'.
Longman's Chemistry film loop no. 8, 'Experiments to show Maxwell-Boltzmann distribution of velocities and energies in molecules of gas'.

Suggestions for homework
Problems 6 and 7.

Summary
Students should know how collision theory can lead to an understanding of experimental kinetic results and what the transition state theory is.

14.5   An investigation of the kinetics of another reaction

Objectives
To investigate the way in which the rate of this reaction varies with time, and to introduce the idea of autocatalysis.

Timing
Five periods should be sufficient.

Suggested treatment
Students met the reaction:

$$2MnO_4^-(aq) + 16H^+(aq) + 5C_2O_4^{2-}(aq) \rightarrow 2Mn^{2+}(aq) + 8H_2O(l) + 10CO_2(g)$$

in the introduction to this topic. In this section they follow the extent of the reaction with time by measuring the concentration of permanganate ions at suitable intervals using a colorimeter.

Experiment 14.5

To investigate the kinetics of the reaction between permanganate and oxalate ions

Each pair of students will need:

1 stopclock, or sight of a large clock with seconds hand
1 test-tube
1 colorimeter with test-tube

access to communal burettes containing:

Solution which is 0.1M with respect to oxalic acid and 1.2M with respect to sulphuric acid
0.02M potassium permanganate
0.02M manganese(II) solution
Carbon dioxide generator

The concentration of permanganate may need adjustment for the particular colorimeter used.* The oxalate ion concentration may require modification depending on the permanganate concentration chosen, so that the experiment takes about four or five minutes.

Procedure

Having adjusted the meter to maximum with a tube of water in place, students mix the oxalic acid and permanganate solutions and take readings with this solution in place. The reaction takes about four minutes, at the end of which time the permanganate will have been consumed. Students are instructed to start by taking readings every thirty seconds but after two minutes, when the rate of the reaction increases, they will need to take readings more frequently, possibly at ten second intervals. If they do not obtain all the readings they require, the experiment can easily be repeated as it only takes three or four minutes. The colorimeter readings are then converted to molarities and a concentration against time graph drawn. This will be of the form shown in figure 14.5 opposite.

Teachers may think it useful for students to take gradients from this graph, and to plot the rates obtained in this way against time. As the concentrations of permanganate, oxalate, and hydrogen ions upon which one might expect the rate to depend are decreasing with time, the rate of the reaction must depend upon the concentration of something which is *increasing* in concentration with time, that is, something which is produced in the reaction. Students will probably need help before they suggest that catalysis by something produced in the reaction is occurring.

* For details of the colorimeter, see Appendix 2.

Looking at the stoichiometric equation, there are two obvious candidates, $Mn^{2+}(aq)$ and $HCO_3^-(aq)$ from dissolved carbon dioxide. Other possibilities are intermediates that may be produced in the reaction such as manganese in other oxidation states. Students can investigate these possibilities by repeating the experiment, adding 1, 2, 4, or 6 drops (approximately 0.02, 0.04, 0.08, or 0.12 cm^3) of 0.2M manganese(II) solution or by saturating the oxalic acid solution with carbon dioxide before the experiment. They will find that the addition of manganese(II) moves the $[MnO_4^-]$ against time graph to the left, that is, moving towards a more 'normal' graph. If enough manganese(II) is added the increase in rate will not occur at all. This indicates that manganese(II) is catalysing the reaction.

Students can then be asked how they think the rate of the reaction between iodine and acetone varies with time, if one starts with acetone and iodide ions only. As hydrogen ions are produced, it is also autocatalytic and it would be expected that the form of the extent of reaction/time graph would be similar to that formed for the permanganate–oxalate reaction.

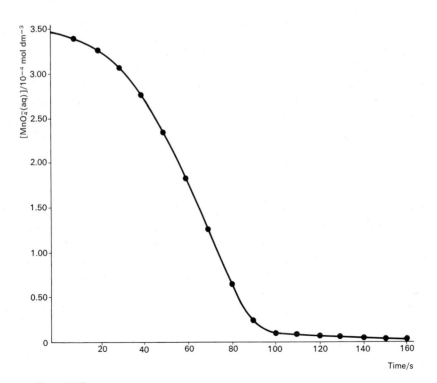

Figure 14.5

Students will want to put forward ideas about the mechanism of the reaction. It is obvious that the reduction of permanganate probably occurs in several steps. The kinetic evidence we have found will not elucidate very much of a proposed mechanism. All we know is that a reaction involving manganese ions provides a faster route than one not involving manganese ions. From this and further evidence a suggested route is:

$$Mn(VII) \xrightarrow{slow} Mn(IV)$$

$$Mn(IV) \xrightarrow{fast} Mn(II)$$

both steps involving the oxidation of the oxalate. When Mn(II) is present

$$Mn(II) + Mn(VII) \xrightarrow{fast} Mn(IV)$$

occurs which does not involve oxidation of the oxalate, followed by

$$Mn(IV) \xrightarrow{fast} Mn(II)$$

which oxidizes the oxalate. This proposed mechanism is a simple and crude one and the first criticism one might make is that the reduction of manganese(VII) to manganese(IV) will probably occur in more than one step. Nevertheless, it is a starting point in trying to explain the kinetics of this reaction and it should be pointed out to students that more evidence is required before one can go further.

The teacher (and students) may wish to spend more time studying this reaction. An article by Dr G. Van Praagh, 'The reaction between permanganate and oxalic acid', in *School Science Review*, (1941) 88, 368 gives suitable experimental instructions for further work.

Alternative method
As an alternative, the reaction can be stopped after a suitable time by adding potassium iodide solution and then the iodine liberated can be titrated. The titre is proportional to the concentration of permanganate at that time. Different members of the class could stop the reaction at different times and use all the results to plot a time/concentration curve. It is best to use 10 cm^3 of the oxalic acid solution and 2 cm^3 of the permanganate solution followed by about 5 cm^3 of approximately 0.1M potassium iodide solution and titration with 0.02M sodium thiosulphate solution. The main difficulty of this method is stopping the reaction accurately.

Summary

In this section, students should have had the opportunity to use some of the ideas of kinetics which they have met, to start to investigate a reaction.

By the end of Topic 14, students should understand what factors affect the rate of a reaction, what is meant by 'the rate expression', how kinetic studies can enable us to set up a model of the mechanism on which the reaction occurs, and something of the theories of chemical kinetics.

Topic 15
Equilibria: redox and acid–base systems

Objective

To use the ideas developed in Topic 12, Equilibria: gaseous and ionic, to study redox equilibria and acid-base equilibria.

Content

15.1   Redox equilibria: metal/metal ion systems.
15.2   Use of e.m.f. measurements to estimate small concentrations of ions.
15.3   Redox equilibria extended to other systems.
15.4   Acid–base equilibria.
15.5   Buffer solutions and indicators.

Timing

About four weeks.

15.1   Redox equilibria: metal/metal ion systems

Objectives

1   To consider the importance to metal activity of relative tendency to form ions in solution.

2   To deal with metal/metal ion systems as redox equilibria.

3   To introduce the hydrogen electrode.

4   To establish the technique for measuring the e.m.f. of cells and to discuss the information obtainable from such measurements.

5   To examine the effects of concentration changes and temperature changes on redox potential values.

6   To define more precisely what is meant by a standard redox potential.

Timing

About eight periods.

Suggested treatment

Much will depend here on the students' previous experience. Those who have followed the Nuffield Chemistry Sample Scheme before starting this course will be familiar with some of the matters dealt with in this section, which appear in Topic 23 of *The Basic Course* and in Option 5 of *A Course of Options*. In these circumstances the treatment suggested below will need to be modified and the time needed will be reduced accordingly.

Introduction

Redox systems have been studied earlier in the course, especially in Topic 5, where a change of oxidation number was used as a criterion for redox processes. In this section we need to concentrate attention on electron transfers which take place in redox reactions. It is suggested that a simple class experiment on two displacement reactions

$$Zn(s) + Cu^{2+}(aq) \rightarrow Zn^{2+}(aq) + Cu(s)$$

and $$Cu(s) + 2Ag^{+}(aq) \rightarrow Cu^{2+}(aq) + 2Ag(s)$$

be used to provoke discussion.

Experiment 15.1a

Some simple redox reactions

Each student or pair of students will need:

4 test-tubes, 100×16 mm
Copper foil, 12×1 cm
Copper powder, about 0.5 g
Zinc foil, 12×1 cm
Zinc powder, about 0.5 g
0.5M copper sulphate solution
0.1M silver nitrate solution
Thermometer, -10 to $+110°C$, $\times 1°$

Procedure

Details are given in the *Students' Book*. Results to be expected are as follows.

1   Zinc foil dipped in copper(II) sulphate solution will acquire a coating of metallic copper.

$$Zn(s) + Cu^{2+}(aq) \rightarrow Zn^{2+}(aq) + Cu(s)$$

2   Zinc powder reacts more quickly with the copper(II) solution, as a greater surface area of the metal is in contact with the solution.

3 and **4** Similar results are obtained with copper and silver nitrate solution:

$$Cu(s) + 2Ag^{+}(aq) \rightarrow 2Ag(s) + Cu^{2+}(aq)$$

Heat is evolved during the reactions, indicating that the energy content of the products is less than that of the reactants in each case.

In (1) and (2)

Zn(o) is oxidized to Zn(II)
and Cu(II) is reduced to Cu(o)

In (3) and (4)

Cu(o) is oxidized to Cu(II)
and Ag(I) is reduced to Ag(o)

Discussion of the questions posed at the end of the experiment should lead to the establishment of the following points.

a   Redox processes can involve transfer of electrons. Loss of electrons is oxidation; gain of electrons is reduction.

b   The terms oxidant (or oxidizing agent) and reductant (reducing agent) are relative only. In the experiment copper is seen to be capable of acting in both senses.

c   The equation for a redox process can be split into 'half-equations' so that the electron transfer is made plain, e.g.

$$Zn(s) \rightarrow Zn^{2+}(aq) + 2e^- \quad oxidation$$
$$2e^- + Cu^{2+}(aq) \rightarrow Cu(s) \quad reduction$$

As these processes are seen to be reversible, they can be treated as equilibria

$$Zn(s) \rightleftharpoons Zn^{2+}(aq) + 2e^-$$
$$\text{and} \quad 2e^- + Cu^{2+}(aq) \rightleftharpoons Cu(s)$$

Application of Le Chatelier's principle to such equilibria tells us that the relative tendency of the two metals to form ions in solution determines the outcome of a given reaction.

d   During a metal/metal ion reaction energy is transferred from system to environment.

e   It is very important at this stage to make it clear that the energy transfer during the change:

$$Zn(s) \rightarrow Zn^{2+}(aq) + 2e^-$$

shown above is *not* the same as the ionization energy of zinc:

$$Zn(g) \rightarrow Zn^{2+}(g) + 2e^-$$

The formation of ions in solution is thus quite different from the conversion of gaseous atoms of a metal into gaseous ions:

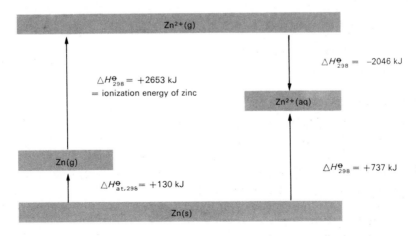

Figure 15.1a

This brief excursion into energetics is intended to do no more than emphasize that ion formation in solution is not a simple process; and it is the tendency to form ions *in solution* that we are concerned with in this part of the course. (It would, of course, be quite wrong to connect $\Delta H^\ominus$ for metal/metal ion reactions with cell e.m.f. values, since the latter are related to free energy factors.)

A point of some interest that may arise in class discussions is that the difference in the value of $\Delta H^\ominus_{298}$ for the change $Zn(s) \rightarrow Zn^{2+}(aq)$ given in figure 15.1a ($+737 \text{ kJ mol}^{-1}$) and that obtained from the *Book of Data* (-152 kJ mol^{-1}) is very large. The reason for this lies in a difference of 'datum level' from which the two values are calculated. An explanation is given below in case it may be needed.

In standard tables of $\Delta H^\ominus_f$ values for ions *in solution* the 'datum level' is $\Delta H^\ominus_f$, $[H^+(aq)] = 0$, an arbitrarily assigned value. For solid, liquid, and gaseous substances the datum level used in standard tables of $\Delta H^\ominus_f$ is that of the elements in their standard states, the values for these being assumed as zero. Thus there are two separate datum levels for these tables.

For figure 15.1a the datum level from which the energy quantities are calculated is $\Delta H^\ominus_{298}$, $Zn(s) = 0$ (element in standard state) and also $\Delta H^\ominus_{298}$, $H_2(g) = 0$. On this basis a value of $+737 \text{ kJ mol}^{-1}$ is obtained for $\Delta H^\ominus_{298}$, $Zn^{2+}(aq)$ and, by a similar thermochemical cycle, of $+953 \text{ kJ mol}^{-1}$ for $\Delta H^\ominus_{298}$, $Cu^{2+}(aq)$. The commonly encountered value of -152 kJ mol^{-1} for $Zn^{2+}(aq)$ can be considered as the energy transfer occurring during the reaction

$$Zn(s) + 2H^+(aq) \rightarrow Zn^{2+}(aq) + H_2(g)$$

assuming that $\Delta H^\ominus_{298}$ is zero for $Zn(s)$, $H^+(aq)$, and $H_2(g)$. A similar assumption allows $\Delta H^\ominus_{298}$ for $Cu^{2+}(aq)$ to be calculated as $+64 \text{ kJ mol}^{-1}$.

This double standard does not cause trouble when $\Delta H^\ominus$ values are being added and subtracted to find enthalpy changes accompanying complete reactions, since the effects of the two datum levels cancel. Thus in the Daniell cell reaction

$$Zn(s) + Cu^{2+}(aq) \rightarrow Zn^{2+}(aq) + Cu(s)$$

we can either use the set of values based on $\Delta H^\ominus$, $H_2(g) = 0$, from which

$$\Delta H^\ominus_{298} = (737 + 0) - (0 + 953) = -216 \text{ kJ}$$

or the more usual values based on $\Delta H^\ominus$, $H^+(aq) = 0$, when

$$\Delta H^\ominus_{298} = (-152 + 0) - (0 + 64) = -216 \text{ kJ}$$

It is when *half-reactions* are under consideration, as in figure 15.1a, that discrepancies arise. The use of the two datum levels, $\Delta H^{\ominus}_{298}$, $H^+(aq) = 0$ and $\Delta H^{\ominus}_{298}$, $H_2(g) = 0$ leads to the erroneous conclusion that for the change

$$H_2(g) \rightarrow 2H^+(aq); \Delta H^{\ominus}_{298} = 0 \text{ kJ}$$

Until fairly recently, it was not possible to arrive at an estimate of this quantity, owing to uncertainty about the enthalpy of hydration of the hydrogen ion

$$H^+(g) \rightarrow H^+(aq)$$

The value of $\Delta H^{\ominus}_{298}$ for this change is concealed within the sequence dealt with in figure 15.1a. A value can be extracted as follows.

Let $\Delta H^{\ominus}_{298}$ for the change $\frac{1}{2}H_2(g) \rightarrow H^+(aq)$ be x kJ and use the value of 737 kJ mol^{-1} for $\Delta H^{\ominus}_{298}$, $Zn^{2+}(aq)$ in the equation

$$Zn(s) + 2H^+(aq) \rightarrow Zn^{2+}(aq) + H_2(g); \Delta H^{\ominus}_{298} = -152 \text{ kJ}$$
$$\quad 0 \qquad 2x \qquad\qquad 737 \qquad\quad 0$$

Then $737 - 2x = -152$

$$\therefore 2x = 737 + 152 = 889$$

and $x = 444$ kJ

Thus for $\frac{1}{2}H_2(g) \rightarrow H^+(aq); \Delta H^{\ominus}_{298} = 444$ kJ

which can be broken down into the following stages:

$$\frac{1}{2}H_2(g) \rightarrow H(g); \Delta H^{\ominus}_{298} = 243 \text{ kJ (atomization energy)}$$

$$H(g) \rightarrow H^+(g) + e^-; \Delta H^{\ominus}_{298} = 1310 \text{ kJ (ionization energy)}$$

$$H^+(g) + aq \rightarrow H^+(aq); \Delta H^{\ominus}_{298} = y \text{ kJ (hydration energy)}$$

$$\frac{1}{2}H_2(g) \rightarrow H^+(aq); \Delta H^{\ominus}_{298} = 243 + 1310 + y \text{ kJ}$$

$$\therefore 243 + 1310 + y = 444$$

and $y = -1109$ kJ

(Reference to three standard texts produced values of -1172, -1084, and -1071 kJ for this quantity.)

The difference between the two values of $\Delta H^{\ominus}_{298}$ for Zn^{2+} (aq) given above, $737 - (-152) = 889$ kJ, is therefore accounted for by the enthalpy change for $H_2(g) \rightarrow 2H^+(aq)$.

Voltaic cells

The possibility of using metal/metal ion reactions as sources of electrical energy in a voltaic cell arises from the half-equations considered in interpreting experiment 15.1a. A voltaic cell provides a means of comparing the relative tendencies of metal/metal ion couples to liberate electrons by forming ions in solution. *Absolute* electrode potentials are difficult to measure but potential *differences* can be measured easily.

Experiment 15.1b explores the variation of potential difference between the electrode systems in a Daniell cell with the change of resistance in a circuit. This enables the idea of the electromotive force (e.m.f.) of a cell as the maximum p.d. obtainable, to be introduced. The symbol E is used for e.m.f. This experiment can be done quickly, as a teacher demonstration. It should be preceded by a brief description of the Daniell cell, including the function of the porous pot in preventing admixture of the electrode solutions whilst allowing electrical contact between them.

Experiment 15.1b

The variation of p.d. in a Daniell cell with change of external resistance

The teacher will need:

Rheostat (1000 ohms or more)
Ammeter (0–1 amp)
Two voltmeters, both 0–2 volt, one of fairly low resistance, and the other of high resistance (preferably a valve voltmeter). The low resistance voltmeter should be in the circuit at the beginning of the experiment.

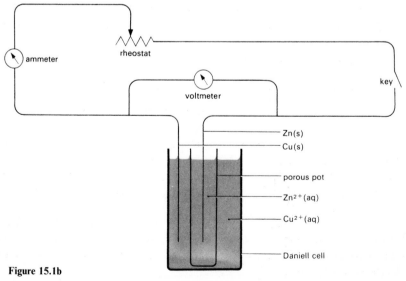

Figure 15.1b

Procedure

1 Close the key, gradually increase the resistance in the cell circuit, and note the effect on ammeter and voltmeter readings.

2 Measure the p.d. with the key open.

3 Replace the low resistance voltmeter with a high resistance instrument and measure the p.d. with the key open.

4 Note the polarity of the cell plates from the voltmeter connection.

The observations show that the p.d. increases to a maximum value as the current falls to zero. The maximum p.d. is called the electromotive force of the cell (E volt).

Taking a voltmeter measurement across the cell terminals, key open, involves a flow of current through the voltmeter, so that an accurate E value cannot be obtained in this way unless the resistance of the voltmeter is so high that the current taken is negligible. A valve (or transistor) voltmeter is the best instrument for this purpose. If one is not available, a potentiometer should be used. Details are given in Appendix 2 in this book, and in the *Students' Book* as Appendix I to the topic.

Cell diagrams

The IUPAC convention is commonly used in Britain and Europe. In this the sign before the E value gives the polarity of the *righthand electrode* in the cell diagram. Of itself, of course, the e.m.f. always represents an energy loss to the system so the $+$ or $-$ sign has no connection with gain or loss of electrical energy. Thus for the Daniell cell, the appropriate method of representation can be either

$$Zn(s)|Zn^{2+}(aq)|Cu^{2+}(aq)|Cu(s); E = +1.1 \text{ V}$$

$$\text{or } Cu(s)|Cu^{2+}(aq)|Zn^{2+}(aq)|Zn(s); E = -1.1 \text{ V}$$

The vertical broken line represents the salt bridge or porous partition.

The contributions made by separate electrode systems to the e.m.f. of a cell

Measurement of the potential of a single electrode system is impossible, because two such systems are needed to make a complete cell of which the e.m.f. can be measured. We can, however, assess the *relative* contributions of single electrode systems to cell e.m.f.s by choosing one system as a standard against which all other systems are measured. The standard system is then arbitrarily assigned zero potential and the potentials of all other systems referred to this value. By international agreement the hydrogen electrode has been chosen as the reference electrode for this purpose.

The hydrogen electrode

This is explained in the *Students' Book*. Although measurements of the e.m.f. of combinations of other electrode systems with the hydrogen electrode sometimes give rise to difficulties, the attempt is worth making as a teacher demonstration. A valve voltmeter is most convenient for this purpose but a potentiometer method can be used quite satisfactorily as an alternative. (See Appendix 2 in this book and Appendix I to this topic in the *Students' Book*.)

Experiment 15.1c

To measure the e.m.f. of simple cells using the hydrogen electrode as the common reference electrode

The teacher will need:

Hydrogen electrode; platinized Pt wire dipping into 1.0M HCl(aq) in small squat-form beaker (50 or 100 cm^3) and hydrogen supply (see Appendix 2)
$Cu^{2+}(aq)|Cu(s)$ electrode
$Zn^{2+}(aq)|Zn(s)$ electrode
$Ag^+(aq)|Ag(s)$ electrode
These electrodes should consist of about 6×1 cm strips of metal foil, cleaned with fine emery paper before use, dipping into solutions of 1M copper(II) sulphate, 1M zinc sulphate, and 0.1M silver nitrate respectively in small squat-form beakers (50 or 100 cm^3). The metal strips can be supported in small clamps between pieces of cork, and electrical connection made to them by leads with a crocodile clip attached to one end. A copper foil backing, kept well clear of the solution, is useful in enabling a good connection to be made between crocodile clip and silver eledtrode.
Potassium nitrate salt bridges. These consist of a single strip of filter paper, 10×1 cm, soaked in saturated potassium nitrate solution.
Valve voltmeter (pH meter with appropriate millivolt scale is suitable). If one is not available, a potentiometer should be used (see Appendix 2 in this Guide and Appendix I to this topic in the *Students' Book*).

Procedure

Set up apparatus as in figure 15.1c. Have the hydrogen electrode on the *lefthand side as the class sees the apparatus*. The end of the glass tube surrounding the platinized platinum wire should be immersed in the 1.0M HCl(aq) as near to the surface as possible, so that the gas pressure is very nearly 1 atm. Use the crocodile clips to connect the electrodes to the valve voltmeter. Note that the copper electrode must be connected to the $+$ terminal of the voltmeter.

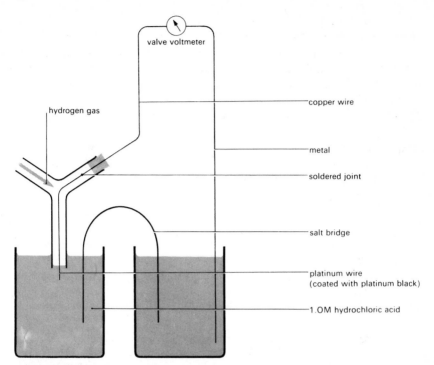

Figure 15.1c

Adjust the flow of hydrogen to a rate of about one bubble every two seconds. Read the cell e.m.f. on the valve voltmeter.

(*Note*. If the platinum black has been exposed to the air, the hydrogen electrode will take a little time to reach equilibrium.)

This e.m.f. will be the E value for the cell:

$$Pt\ [H_2(g)]\,|\,2H^+(aq)\,|\,Cu^{2+}(aq)\,|\,Cu(s)$$

The experiment can then be repeated with zinc and silver foils in place of the copper foil. Note that in the case of zinc, the hydrogen electrode will have to be connected to the $+$ terminal of the voltmeter; in the case of silver, as of copper, the hydrogen electrode will have to be connected to the $-$ terminal. These e.m.f.s will be the E values for the cells:

$$Zn(s)\,|\,Zn^{2+}(aq)\,|\,2H^+(aq)\,|\,[H_2(g)]\ Pt$$

$$Pt\ [H_2(g)]\,|\,2H^+(aq)\,|\,Ag^+(aq)\,|\,Ag(s)$$

Discussion

Results should be as now indicated:

a   Pt $[H_2(g)] | 2H^+(aq) | Cu^{2+}(aq) | Cu(s)$; $E = +0.32$ V
b $Zn(s) | Zn^{2+}(aq) | 2H^+(aq) | [H_2(g)]$ Pt; $E = +0.75$ V
c   Pt $[H_2(g)] | 2H^+(aq)$ $Ag^+(aq) | Ag(s)$; $E = +0.76$ V

The E values obtained give no indication of the absolute potentials that can be attributed to an individual electrode. To do this by a method of this kind we should have to find an electrode with zero potential, in which case there would be no flow of electrons into or out of it.

The simplest way of assessing the relative contributions of single electrode systems to cell e.m.f. values is to choose one electrode system as a reference standard and measure the E values of all other systems against this. The standard chosen is, of course, the hydrogen electrode. Giving the potential of this electrode the value of zero, and rewriting cell (b) in the reverse order we have:

a   Pt $[H_2(g)] | 2H^+(aq) | Cu^{2+}(aq) | Cu(s)$; $E = +0.32$ V
b   Pt $[H_2(g)] | 2H^+(aq) | Zn^{2+}(aq) | Zn(s)$; $E = -0.75$ V
c   Pt $[H_2(g)] | 2H^+(aq) | Ag^+(aq) | Ag(s)$; $E = +0.76$ V

We can represent these values on a linear chart,

$Zn^{2+}(aq) | Zn(g)$	$2H^+(aq) | [H_2(g)]$ Pt	$Cu^{2+}(aq) | Cu(s)$	$Ag^+(aq) | Ag(s)$
-0.75 V	0 V	$+0.32$ V	$+0.76$ V

The redox potential series given in table TEI of the *Book of Data* can now be discussed briefly. The point should be made that these $E^\ominus$ values are obtained under carefully specified conditions (hence the superscript) which will be discussed in the next section. Therefore we should not expect exact comparison with the values that have been obtained experimentally in experiment 15.1c. The most important point to stress at this stage is that the E values given are the voltages of *real and complete* cells, of which the lefthand electrode is always

$Pt_2[H(g)] | 2H^+(aq)$ (Hence its position in experiment 15.1c.)

The order of oxidants and reductants is important in the above chart; it is the order given in the tables of electrode potentials framed according to the IUPAC convention. (It is unfortunate that some countries adopt the opposite convention which entails a reversal of sign for E values, e.g. $+0.75$ for $Zn(s) | Zn^{2+}(aq)$; if old American textbooks are used as reference sources, students should be warned of this discrepancy.)

Experiments 15.1d and 15.1e are intended to give an opportunity for students to gain experience in constructing voltaic cells and making e.m.f. measurements.

Experiment 15.1d
To measure the e.m.f. of some voltaic cells

Each student or pair of students will need:

Copper, zinc and silver half cells as in experiment 15.1c
Potassium nitrate salt bridges as in experiment 15.1c

access to:

1 or preferably 2 valve voltmeters. If these are not available, each student or pair of students will require a potentiometer. Details of such an instrument are given in Appendix 2 of this Guide, and in Appendix I to this topic in the *Students' Book*.

Details of alternative cells

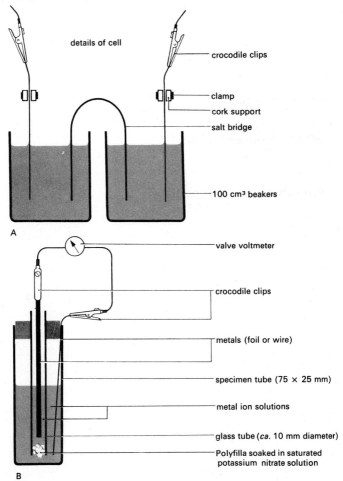

Figure 15.1d

A further alternative type of cell is described in Nuffield Chemistry *The Sample Scheme Stage III, A Course of Options*, on page 97; use of this version, and type B in figure 15.1d, has two advantages:

1   There is a considerable saving of chemicals, especially if the silver nitrate solution is placed in the inner tube.

2   The complete cell is easily portable, hence each student (or pair) can make up one cell, which can be shared with other students.

Procedure

Students measure the e.m.f.s of the cells:

a $Cu(s)|Cu^{2+}(aq)|Zn^{2+}(aq)|Zn(s)$

b $Ag(s)|Ag^+(aq)|Cu^{2+}(aq)|Cu(s)$

c $Ag(s)|Ag^+(aq)|Zn^{2+}(aq)|Zn(s)$

The salt bridge must be changed for each measurement. Full details are given in the *Students' Book*.

Discussion

Results obtained by a sixth form set in this experiment are given below. They are class averages, omitting one result for cell (a) which was obviously incorrect.

a $Cu(s)|Cu^{2+}(aq)|Zn^{2+}(aq)|Zn(s)$; $E = -1.07$ V

b $Ag(s)|Ag^+(aq)|Cu^{2+}(aq)|Cu(s)$; $E = -0.44$ V

c $Ag(s)|Ag^+(aq)\ Zn^{2+}(aq)|Zn(s)$; $E = -1.53$ V

Answers to the questions posed in the *Students' Book* should be along the following lines.

1   Comparison of these results with the displacement reactions in experiment 15.1a shows that the electrode system with the greater tendency to form ions is the negative pole in each cell.

2   The E value for cell (c) is very nearly that obtained by adding the E value for cell (a) to that for cell (b). This is reasonable because by connecting cells (a) and (b) together as written below

$$Ag(s)|Ag^+(aq)|Cu^{2+}(aq)|\overbrace{Cu(s)\quad Cu(s)}|Cu^{2+}(aq)|Zn^{2+}(aq)|Zn(s)$$
$$\quad\quad\quad -0.44\text{ V}\quad\quad\quad\quad\quad\quad\quad -1.07\text{ V}$$

we should expect the effects of the two copper electrodes to cancel since they oppose each other and no electron flow should take place between them.

3   The e.m.f.s calculated from the list given in the *Students' Book* are:

a $Cu(s)|Cu^{2+}(aq)|H^+(aq)|[H_2(g)]$ Pt; $E = -0.34$ V

b   Pt $[H_2(g)]|H^+(aq)|Zn^{2+}(aq)|Zn(s)$; $E = -0.76$ V

Therefore

$$Cu(s)|Cu^{2+}(aq)|H^+(aq)|[H_2(g)] \overset{\frown}{Pt} \quad Pt\ [H_2(g)]|H^+(aq)|Zn^{2+}(aq)|Zn(s)$$

in which the two hydrogen electrodes cancel out, will have an e.m.f.

$$-0.34 + (-0.76)\ V$$
$$= -1.1\ V$$

similarly (b) should have an e.m.f. of -0.46 V
and   (c) should have an e.m.f. of -1.56 V

Experiment 15.1e
To investigate the effect of changes in silver ion concentration on the potential of the $Ag^+(aq)|Ag(s)$ electrode

Each student or pair of students will need:

Copper and silver half cells as in experiment 15.1c. (Solutions required are listed below.)
Potassium nitrate salt bridges as in experiment 15.1c.
[If cell B is used (given in *A Course of Options*) put $Ag^+(aq)$ in the inner tube: cells of different concentrations can be made up by different students (or pairs) and interchanged for measurements.]

access to:

Valve voltmeters or alternatives as in experiment 15.1d
1.0M copper sulphate solution, about 40 or 70 cm^3 depending on beaker size used
Silver nitrate solution in concentrations of 0.01M, 0.0033M, 0.001M, 0.00033M, and 0.0001M*
   Time will be saved if these are prepared beforehand. Otherwise students will need 10 cm^3 pipettes and 100 cm^3 standard flasks to dilute 0.01M solution. 0.001M and 0.0001M are prepared by progressive dilution. 0.0033M by mixing one volume 0.01M with two volumes water; 0.00033M is prepared similarly from 0.001M.
The cell is set up as indicated in figure 15.1c.

Procedure
Using the principle of Le Chatelier the students are asked to predict the effect of concentration change on the equilibrium

$$Ag(s) \rightleftharpoons Ag^+(aq) + e^-$$

Decrease of silver ion concentration increases the tendency for metallic silver to form ions in solution. It has been found that the electrode system in a cell with the greater tendency to form ions is the negative pole in each cell, that is, it has the more negative or less positive electrode potential. Thus, if the tendency for the metal to form ions is increased, the electrode potential would be expected to be more negative or less positive. A similar result would be expected for other metal/metal ion equilibria. This prediction is then tested.

* The activity coefficient of the most concentrated of these solutions, 10^{-2}M, is 0.90; those of the other concentrations are larger. Thus lg $[Ag^+(aq)]$ is very close to lg $a_{Ag^+(aq)}$. For the 10^{-2}M solution lg $[Ag^+(aq)] = -2.00$ and lg $a_{Ag^+(aq)} = -2.05$.

The cell

$$Cu(s)|Cu^{2+}(aq) \ Ag^+(aq)|Ag(s)$$

is set up as indicated in figure 15.1d and the e.m.f. of the cell measured for various values of $[Ag^+(aq)]$. Instructions are given in the *Students' Book*.

The results for this experiment of variation of $Ag^+(aq)$ concentration will be needed for measurement of silver concentrations in experiment 15.2a. If valve voltmeters are not available, it is advisable, therefore, for students to calibrate the potentiometer wire by inserting a standard cell in place of the test cell, at the beginning and end of the experiment. A Weston standard cell ($E = 1.018$ V) is conventionally used for this but a Mallory mercury cell ($E = 1.35$ V) can provide an adequate substitute.

Results are tabulated as $[\text{ion}]/\text{mol dm}^{-3}$; $\lg[\text{ion}]$; E/volt.

Graphs of E against $\lg[\text{ion}]$ are drawn. An extended concentration axis permits extrapolation to $\lg[Ag^+(aq)] = -18$ for use later. Students may need some help in plotting this graph. The e.m.f.s are all positive and the values of $\lg[\text{ion}]$ negative.

In order to find the value of $\lg[Ag^+(aq)]$ when $E = 0$ (and the value of E when $\lg[Ag^+(aq)] = -18$) the graph will be as shown in figure 15.1e.

Answers to the questions posed in the *Students' Book* should be along the following lines.

1   The e.m.f. for the reaction is made up of the two electrode potentials

$$Cu(s)|Cu^{2+}(aq)|2H^+(aq)|[H_2(g)] \ Pt \quad Pt[H_2(g)]|2H^+(aq)|Ag^+(aq)|Ag(s)$$
$$-(+0.34 \text{ V}) \qquad\qquad\qquad +0.80 \text{ V}$$
$$\text{e.m.f.} = -0.34+0.80 = +0.46 \text{ V}$$

Decrease of silver ion concentration will increase the tendency of metallic silver to form ions in solution and therefore the potential of the silver electrode would be expected to become more negative or less positive. The value of the cell e.m.f. will therefore become more negative or less positive. This is shown by the graph in figure 15.1e.

2   As the graph is a straight line

$$E \propto \lg[\text{ion}]$$

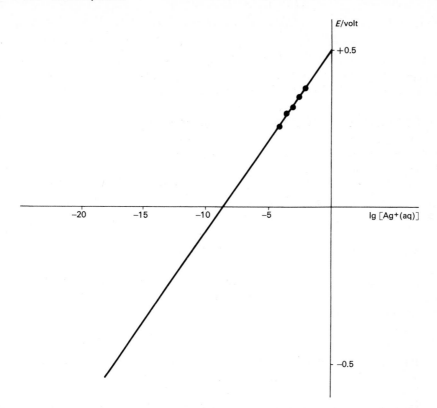

Figure 15.1e

3   Using the graph of silver ion concentration against E, the concentration of silver ions for which $E = 0$ can be found. In the cell used $[Cu^{2+}(aq)] = 1$ mol dm^{-3}. This gives the information required to calculate the equilibrium constant for the reaction

$$Cu(s) + 2Ag^+(aq) \rightleftharpoons Cu^{2+}(aq) + 2Ag(s).$$

From a class experiment the value of $lg\,[Ag^+(aq)]$ for $E = 0$ was found, by extrapolation, to be -8.8.

$$\therefore\ [Ag^+(aq)] = antilg\,\bar{9}.2 = 1.6 \times 10^{-9}$$

$$\therefore\ \qquad K_c = \frac{[Cu^{2+}(aq)]_{eqm}}{[Ag^+(aq)]^2_{eqm}} = \frac{1}{(1.6 \times 10^{-9})^2} = 4 \times 10^{17}\ dm^3\ mol^{-1}$$

These experiments show that the value of an electrode potential depends on the concentration of ions in solution. It can also be shown that E varies with temperature. Thus for comparison purposes the temperature and concentration must be standardized. A more precise definition of *standard redox potential* now becomes possible. For our purposes we shall define it as the e.m.f. of a cell in which one electrode is the standard hydrogen electrode and the other electrode consists of a metal in contact with a molar solution of its ions, the e.m.f. being measured at 25 °C. A standard hydrogen electrode consists of hydrogen gas, at a pressure of one atmosphere, bubbling over a platinized platinum surface in a solution which is molar with respect to hydrogen ions.

Standard redox potentials are denoted by the symbol $E^{\ominus}$ ('E standard' when spoken). A list is given in table TEI of the *Book of Data*.

A minute or two could be spent in pointing out that the concentration of free ions cannot strictly be equated with the solute concentration as calculated by number of moles of solute in a given volume of solution. Incomplete ionization, inter-ion attraction, and 'crowding effects' in fairly concentrated solution result in the necessity of taking a greater concentration of solute to achieve a given concentration of ions. Thus to obtain a 1.0M solution of free hydrogen ions a concentration of 1.18M hydrogen chloride must be used. No account will be taken of activities or activity coefficients in this topic but students should be aware that the treatment given here is simplified to some extent.

The information that has been collected so far can be summarized as follows:
   1   The standard redox potential ($E^{\ominus}$) of an electrode is an intensive property and has a specific value for each electrode system.
   2 $E^{\ominus}$ values are relative, the standard hydrogen electrode being given the arbitrary value of zero. The sign of $E^{\ominus}$ is the polarity of the metal/metal ion electrode when combined with the standard hydrogen electrode; if it forms the negative pole, the $E^{\ominus}$ value is given a negative sign, and vice versa.
   3 When the ion concentration is other than molar, the E value (not the $E^{\ominus}$ value) varies with concentration as predicted by the principle of Le Chatelier. E is directly proportional to lg [ion]. The size of the variation depends on the metal used.
   4 The E value is a function of temperature.

The Nernst equation
At temperatures fairly near to 25 °C the variation of E with concentration changes for metal/metal ion systems can be calculated with a fair approximation to accuracy from the relationship

$$E = E^{\ominus} + \frac{0.06}{z} \lg [\text{ion}]$$

where z = number of charges on metal ion (change in oxidation number).

(The constant has the value of 0.057 at 15 °C; 0.059 at 25 °C; 0.060 at 30 °C.) This fits in with the Le Chatelier prediction of concentration effects since for values of [ion] which are less than 1, lg [ion] is negative and the value of E becomes less positive (more negative) with dilution.

The use of graphs obtained in experiment 15.1e to test this relationship is explained in the *Students' Book*. For the silver ion the slope of the E against lg [ion] line, which should be 0.06, usually gives quite a good agreement.

The full Nernst equation

$$E = E^{\ominus} + \frac{RT}{zF} \ln \text{[ion]}$$

for metal/metal ion systems is given in the *Students' Book* with a brief explanation of the terms involved. Less able students may make heavy weather of this, so it should not be stressed too much. In case the point is raised, it should be borne in mind that the equation cannot be used to calculate E values at temperatures other than 25 °C, since $E^{\ominus}$ varies with temperature also. The temperature term relates solely to corrections for varying ion concentration.

By making use of known variation of ion concentration with E, the concentration of ions in very dilute solutions can be measured electrically. This method has drawbacks with concentrated solutions but can be used for solutions which would be much too dilute to be analysed by ordinary chemical methods. Some examples of the method are contained in the next section.

Suggestions for homework
Calculating cell e.m.f. for different electrode combinations from information in table TEI of the *Book of Data*. Writing equations for cell reactions. Identifying positive and negative electrodes.

Summary
At the end of this section students should:

1   Have some familiarity with the electron transfer aspect of redox reactions, and be able to divide these into half-reactions in simple cases.

2   Appreciate that redox reactions can involve equilibria which are governed by the relative tendencies of the half reactions to liberate electrons.

3   Appreciate the fundamental differences between ionization of gaseous atoms, and ion formation in solution from crystal lattices.

4   Understand why metal/metal ion reactions can be used as energy sources in voltaic cells.

5   Be able to interpret simple cell diagrams.

6   Realize the basis of the conventional tables of redox potentials.

7   Have a clear idea of what is meant by a *standard* redox potential of a metal/metal ion system.

15.2   Use of e.m.f. measurements to estimate small concentrations of ions

Objective

To use the relationship between silver ion concentration and E value for the silver electrode to explore solubility differences for sparingly soluble silver salts.

Timing

About three periods.

Suggested treatment

The extended graph of lg $[Ag^+(aq)]$ against E obtained from the measurements in experiment 15.1e can be used to estimate values for silver ion concentrations much smaller than would be possible by conventional analytical methods. Some examples of this are given in experiment 15.2a using the copper electrode with 1.0M $Cu^{2+}(aq)$ as a reference electrode in the cell

$$Cu(s)|Cu^{2+}(aq)|Ag^+(aq)|Ag(s)$$

Mixtures of 0.1M silver nitrate solution with various precipitating reagents are used in the silver electrode.

The four parts of the experiment can be shared out amongst members of the class if necessary. The experiment should not occupy more than one double period.

Experiment 15.2a

To investigate the changes in silver ion concentration in solution when different reagents are added

Each pair of students will need:

Copper and silver half cells as in experiment 15.1c. (Solutions required are listed below.)
Potassium nitrate salt bridges as in experiment 15.1c
(The alternative cells suggested on page 93 can be used also; put the silver solution in the inner tube)
Measuring cylinder, 25 cm^3

access to:

Valve voltmeters or alternatives as in experiment 15.1d

the following solutions in approximately the volumes stated:

1.0M $CuSO_4$(aq), 40 or 70 cm³, depending on beaker size used
0.1M $AgNO_3$(aq), 30 cm³
0.1M KCl(aq) or NaCl(aq), 30 cm³ ⎫
0.1M KBr(aq), 30 cm³   ⎪
0.1M KI(aq), 30 cm³   ⎬ according to allocated reaction
0.1M KIO_3(aq), 30 cm³   ⎭

(*Note.* For the alternative types of cell much smaller volumes of reagents are adequate.)

Procedure

This is described in the *Students' Book*. For most of the mixtures used in the silver electrode system, the reaction taking place reduces the silver ion concentration to such an extent that the sign of E is reversed relative to the reference copper electrode system. Students should be warned of this possibility and told to deal with it by reversing the electrode connections. If this is done, the silver electrode now becomes the *negative pole* of the cell and measured E values must be adjusted accordingly.

From the E values measured for the various electrode combinations, the corresponding values of lg $[Ag^+(aq)]$ are obtained from the graph. Values of $[Ag^+(aq)]$ are then calculated. Some help may be needed when students change from negative lg value to negative (bar) characteristics and positive mantissa, for example, converting lg $[Ag^+(aq)] = -5.4$ to $\bar{6}.6$, so that, from tables, $[Ag^+(aq)] = 4 \times 10^{-6}$. No more than one significant figure should be used to express the final results. The method of calculating solubility product values from the results obtained is given in the *Students' Book*.

Discussion

The results obtained by the whole class should be collected together and used to discuss briefly the relative solubility of the silver compounds studied (AgCl, AgBr, AgI, $AgIO_3$). The value for the solubility product of silver iodate found by the electrical method should be compared with that obtained by titration in experiment 12.3.

Some reasonably accurate values for solubility products are given in table CHM of the *Book of Data* and the relation between K_{sp} and solubility is illustrated by two worked examples in the *Students' Book*.

The test-tube experiments in experiment 15.2b are intended to enable students to test predictions they make using solubility products.

Note. For the information of the teacher, it is not possible in practice to realize the extended graph of E against lg $[Ag^+(aq)]$ used in experiment 15.2a. If the dilution of silver nitrate solution is continued beyond the limit suggested in the experiment (10^{-4}M) the cell e.m.f. becomes constant in the way shown by the following results (obtained using a potentiometer):

$[Ag^+(aq)]$/mol dm^{-3}	10^{-1}	10^{-2}	10^{-3}	10^{-4}	10^{-5}	10^{-6}	10^{-7}	10^{-8}	10^{-9}
E/volt	0.41	0.37	0.31	0.25	0.20	0.19	0.17	0.18	0.17

It seems likely that the limiting value for E in this system is determined by the equilibrium

$$Ag^+(aq) + OH^-(aq) \rightleftharpoons AgOH(s)$$

since, when distilled water was used instead of $AgNO_3(aq)$ in the silver electrode system, the measured cell e.m.f. was 0.17 V, corresponding to a silver ion concentration of 3×10^{-6} mol dm^{-3}. This does not, of course, invalidate concentration measurements smaller than this value but in order to obtain lower values for $[Ag^+(aq)]$ a different system must be used. If $I^-(aq)$ ions are added to an electrode system which contains $OH^-(aq)$ ions and $Ag^+(aq)$ ions, the electrode potential of the system is determined by the equilibrium

$$Ag^+(aq) + I^-(aq) \rightleftharpoons AgI(s)$$

for which K_{sp} is much smaller than for the $Ag^+(aq)/OH^-(aq)$ equilibrium. A cell consisting of a standard copper electrode combined with a silver electrode dipping into potassium iodide solution gives a negative e.m.f. value considered from the cell diagram

$$Cu(s) \mid Cu^{2+}(aq) \vdots Ag^+(aq) \mid Ag(s),$$

of the same order as that given by the mixture of KI(aq) and $AgNO_3(aq)$ in experiment 15.2a. Some results for a range of electrolytes in the negative electrode are shown in table 15.2.

Electrolyte in Ag electrode (ratios by volume)	E/volt	lg $[Ag^+(aq)]$ (graph)	$[Ag^+(aq)]$/mol dm^{-3}	$[I^-(aq)]$/mol dm^{-3}	K_{sp}/mol^2 dm^{-6}
0.1M $Ag^+(aq)$/ 0.1M $I^-(aq)$ = 2:3	-0.36	-14.8	1.6×10^{-15}	2×10^{-2}	3×10^{-17}
0.1M $Ag^+(aq)$/ 0.1M $I^-(aq)$ = 1:2	-0.37	-15	10^{-15}	3.3×10^{-2}	3×10^{-17}
0.1M $Ag^+(aq)$/ 0.1M $I^-(aq)$ = 3:7	-0.38	-15.2	6.3×10^{-16}	4×10^{-2}	3×10^{-17}
0.1M $I^-(aq)$	-0.41	-15.8	1.6×10^{-16}	10^{-1}	2×10^{-17}

Table 15.2

Experiment 15.2b
A revision exercise

Each student will need:

3 test-tubes
Dropping pipette

access to approximately 0.1M *solutions of:*

silver nitrate
potassium chloride
potassium iodide
potassium bromide
potassium chromate
potassium iodate

Procedure

1   Silver nitrate solution is added drop by drop to a solution containing chloride and iodide ions. As the solubility product of silver iodide (8×10^{-17} mol^2 dm^{-6}) is much smaller than that of silver chloride (2×10^{-10} mol^2 dm^{-6}), silver iodide will precipitate first and, until nearly all the iodide ions have been removed from the solution by precipitation, the silver ion concentration will not rise sufficiently for silver chloride to be precipitated.

2   Solid silver chloride is shaken with potassium iodide solution. As the solubility product of silver iodide is much smaller than that of silver chloride, the equilibrium:

$$Ag^+(aq) + Cl^-(aq) \rightleftharpoons AgCl(s)$$

is disturbed to the left by removal of silver ions. Thus silver chloride dissolves and silver iodide is precipitated.

3   Solid silver chromate is shaken with potassium chloride solution. For the equilibrium:

$$Ag_2CrO_4(s) \rightleftharpoons 2Ag^+(aq) + CrO_4^{2-}(aq)$$

$$[Ag^+(aq)]^2_{eqm}[CrO_4^{2-}]_{eqm} = 3 \times 10^{-12} \ mol^3 \ dm^{-9}$$

$$\therefore [Ag^+(aq)]_{eqm} = \sqrt[3]{3 \times 10^{-12}} \ mol \ dm^{-3}$$

$$= 1.3 \times 10^{-4} \ mol \ dm^{-3}$$

This is large enough to precipitate AgCl, as, for the equilibrium

$$AgCl(s) = Ag^+(aq) + Cl^-(aq),$$

$$[Ag^+(aq)]_{eqm}[Cl^-(aq)]_{eqm} = 2 \times 10^{-10} \ mol^2 \ dm^{-6}$$

$$\therefore [Ag^+(aq)]_{eqm} = \sqrt[2]{2 \times 10^{-10}}$$

$$= 1.4 \times 10^{-5} \ mol \ dm^{-3}$$

So silver chromate dissolves as the equilibrium is disturbed by the precipitation of silver chloride.

4   Solid silver iodate is shaken with potassium bromide solution. The solubility product of silver bromide is much smaller than that of silver iodate, the equilibrium

$$Ag^+(aq) + IO_3^-(aq) \rightleftharpoons AgIO_3(s)$$

is disturbed to the left by removal of silver ions. Silver iodate dissolves and silver bromide is precipitated.

Suggestions for homework
Calculating solubility products and solubilities of sparingly soluble compounds.

Summary
At the end of this section students should appreciate how electrical methods can be used to estimate very small concentrations of metal ions, and how such methods can be used to compare the solubilities of sparingly soluble compounds.

15.3   Redox equilibria extended to other systems

Objectives
1   To study equilibria involving ion/ion and non-metal/non-metal ion reactions, and to use these in voltaic cells.

2   To investigate the effect of concentration changes on E values for electrode systems based on ion/ion reactions.

Timing
About three periods.

Suggested treatment
For this treatment OP transparency number 120 will be useful.

So far in this topic we have been concerned with redox systems involving a metal in equilibrium with its ions in solution. Experiment 15.3a introduces other types of redox reaction by a brief qualitative study of the reaction

$$2Fe^{3+}(aq) + 2I^-(aq) \rightarrow 2Fe^{2+}(aq) + I_2(aq)$$

Experiment 15.3a

To investigate the reaction between iron(III) ions and iodide ions

Each student will need:

Test-tubes and rack

access to solutions of the following materials, of approximately the concentrations indicated:

0.1M Fe^{3+}(aq); iron(III) sulphate or ammonium iron(III) sulphate (iron alum) are suitable
0.1M potassium iodide
0.1M Fe^{2+}(aq); iron(II) sulphate is suitable
1 per cent starch
2 per cent potassium hexacyanoferrate(III), $K_3Fe(CN)_6$

Procedure

Details of the procedure are given in the *Students' Book*. Formation of a deep blue colour with potassium hexacyanoferrate(III) solution and with starch solution indicates that iron(II) and iodine are formed when solutions of iron(III) and iodide ions are mixed.

Use of the iron(III)/iodide reaction in a voltaic cell. The electron transfer which occurs during this reaction can be seen from the half-reactions

$$Fe^{3+}(aq) + e^- \rightarrow Fe^{2+}(aq) \ reduction$$
and $\quad 2I^-(aq) \rightarrow I_2(aq) + 2e^- \quad oxidation$

which arise from the competing equilibria

$$Fe^{3+}(aq) + e^- \rightleftharpoons Fe^{2+}(aq)$$
and $\quad 2I^-(aq) \rightleftharpoons I_2(aq) + 2e^-$

The qualitative study enables the equilibrium position for the complete reaction

$$2Fe^{3+}(aq) + 2I^-(aq) \rightleftharpoons 2Fe^{2+}(aq) + I_2(aq)$$

to be established as lying towards the righthand side of the equation.

The use of this reaction in a voltaic cell is studied in experiment 15.3b.

Experiment 15.3b

To measure the redox potentials for the $Fe^{3+}(aq)/Fe^{2+}(aq)$ equilibrium and the $2I^-(aq)/I_2(aq)$ equilibrium

Each pair of students will need:

Copper reference electrode, as in experiment 15.1c ⎫
Potassium nitrate salt bridges, as in experiment 15.1c ⎪ or alternative versions
2 smooth platinum electrodes   ⎬ suggested in experiment 15.1c
3 beakers, 50 cm³ or 100 cm³   ⎭

access to:

Arbitrary solutions of:
  iodine in aqueous KI
  a mixture of iron(II) and iron(III) salts
Valve voltmeters or alternatives as in experiment 15.1d
The circuit is shown in figure 15.3.

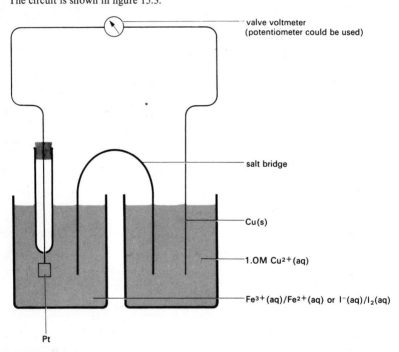

valve voltmeter
(potentiometer could be used)

salt bridge

Cu(s)

1.0M $Cu^{2+}(aq)$

$Fe^{3+}(aq)/Fe^{2+}(aq)$ or $I^-(aq)/I_2(aq)$

Pt

Figure 15.3

Procedure

Students measure the e.m.f.s of the cells:

$$Cu(s)|Cu^{2+}(aq)|Fe^{3+}(aq), Fe^{2+}(aq)|Pt$$
and $$Cu(s)|Cu^{2+}(aq)|I_2(aq), 2I^-(aq)|Pt$$

and from the results calculate the e.m.f. of the cell

$$Pt|2I^-(aq), I_2(aq)|Fe^{3+}(aq), Fe^{2+}(aq)|Pt$$

They check their calculation by measuring the e.m.f. of this cell. A fresh potassium nitrate salt bridge should be used for each measurement.

Discussion

Students' attention should be drawn to the fact that it is necessary to have both the oxidized and reduced forms in each electrode system, to enable the half-cell reactions to proceed in either direction.

The platinum electrode takes no part in the reactions but merely acts as an inert surface by means of which electrons can be transferred into or out of a half-cell.

The agreement between predicted and measured values for the third cell studied shows that ion/ion reactions and non-metal/non-metal ion reactions can be dealt with in the same way as metal/metal ion reactions studied earlier.

To deal with cells of this kind an extension of the conventions for writing cell diagrams is necessary. The reduced form of the redox couple is always placed nearest to the inert electrode and separated from the oxidized form by a comma

$$Fe^{3+}(aq), Fe^{2+}(aq)|Pt$$

oxidized   reduced
form   form

Concentration effects in ion/ion systems

The application of either the principle of Le Chatelier or the equilibrium law to the iron(II)/iron(III) equilibrium indicates that the equilibrium position is affected by relative ion concentrations:

$$Fe^{3+}(aq)+e^- \rightleftharpoons Fe^{2+}(aq)$$

Increase in the relative concentration of $Fe^{3+}(aq)$ drives the equilibrium to the right, reducing the negative potential of the system and making the e.m.f. of the cell more positive.

Temperature also has an effect on the system so that for electrode systems of this kind it is necessary to specify concentration ratios and temperature for the standard redox potential (E). The conditions chosen are:

Equal molar concentrations of reduced and oxidized forms.
A temperature of 298 K (25 °C).

The value of E for other conditions is given by the Nernst equation in the form

$$E = E^\ominus + \frac{RT}{zF} \ln \frac{\text{[oxidized form]}}{\text{[reduced form]}}$$

which gives for a temperature of 25 °C,

$$E = E^{\ominus} + \frac{0.06}{z} \lg \frac{[\text{oxidized form}]}{[\text{reduced form}]}$$

z is the number of electrons transferred when the oxidized form changes to the reduced form.

$$[z = 1 \text{ for } Fe^{3+}(aq) + e^- \rightarrow Fe^{2+}(aq)]$$

Thus, using the values given in table 15.3 (which also appears in the *Students' Book*) a plot of

$$\lg \frac{[Fe^{3+}(aq)]}{[Fe^{2+}(aq)]}$$

against E will give a straight line of slope approximately 0.06.

Relative concentrations/mol dm^{-3}		$\lg \dfrac{[Fe^{3+}]}{[Fe^{2+}]}$	E/volt
$[Fe^{3+}(aq)]$	$[Fe^{2+}(aq)]$		
1	9	-0.954	0.716
2	8	-0.602	0.735
3	7	-0.368	0.748
4	6	-0.177	0.760
5	5	0	0.770
6	4	$+0.177$	0.782
7	3	$+0.368$	0.792
8	2	$+0.602$	0.805
9	1	$+0.954$	0.825

Table 15.3
The variation of redox potential with concentration for the $Fe^{3+}(aq)$, $Fe^{2+}(aq)$ electrode (E values measured against a standard hydrogen electrode)

When [oxidized form] = [reduced form]

$$\lg \frac{[\text{oxidized form}]}{[\text{reduced form}]} = \lg 1 = 0$$

and $E = E^{\ominus}$. This enables a value for $E^{\ominus}$ to be obtained for the iron(III)/iron(II) electrode from the graph by taking the value of E for $\lg[Fe^{3+}(aq)]/[Fe^{2+}(aq)] = 0$.

It should be pointed out that the full Nernst equation

$$E = E^{\ominus} + \frac{RT}{zF} \ln \frac{[\text{oxidized form}]}{[\text{reduced form}]}$$

applies to all equilibria of this kind. In the case of a metal/metal ion system the reduced form is a crystalline metal; the concentration of this is constant so that variations in E depend on [ion] only.

As in preceding sections, attempts to deal more rigorously with the systems under consideration would involve the use of activities. The effects of inter-ionic attraction and incomplete ionization can be mentioned, but no more than this should be attempted.

Suggestions for homework
Calculating cell e.m.f.s from $E^\ominus$ values.

Summary
This section introduced little that is new but rather extends ideas covered in sections 15.1 and 15.2. Use of $E^\ominus$ values for gaining information about voltaic cells is the most important aspect as far as students are concerned.

15.4   Acid–base equilibria

Objectives
1   To introduce the Lowry-Brønsted theory of acids and bases.
2   To consider methods of pH measurement.
3   To use pH measurements for introducing ideas about the strengths of acids and bases.
4   To study pH changes during acid–base reactions as shown by titration curves for strong and weak acids and bases.

Timing
About eight periods.

Suggested treatment
For this treatment OP transparency number 119 will be useful.

This part of the topic can be developed along fairly orthodox lines and has been dealt with in some detail in the *Students' Book*. The text could be read by students before the subject matter is discussed in class.

The initial discussion of acids and bases should be fairly brief and the contrast between competition for protons in acid–base systems and competition for electrons in redox systems serves to link this section with work that has been done previously.

pH values as a means of comparing acidity can then be discussed. A short experimental demonstration using a hydrogen electrode in solutions of differing hydrogen ion concentration is useful here. This can be done with a hydrogen electrode combined with a standard copper electrode with the usual connecting salt bridge and a valve voltmeter for e.m.f. measurement (or a potentiometer if no suitable high resistance voltmeter is available). Students can be asked to predict the effect of changing hydrogen ion concentration on the e.m.f. of the cell used.

$$Pt\,[H_2(g)]\,|\,2H^+(aq)\,|\,Cu^{2+}(aq,\ 1.0M)\,|\,Cu(s)$$

Measurements should be taken for a small range of solutions, such as 0.1M, 0.01M, and 0.001M HCl(aq) and 1.0M and 0.0001M NaOH(aq).

The detection of hydrogen ions in the alkaline solutions is important and can lead to a discussion of the ionization of water and the value of K_w.

All this can be done quickly and the measurements need only be used to indicate trends in equilibrium changes. The hydrogen electrode is not a satisfactory system to use and the idea of the glass electrode and the pH meter should be introduced quite early. If a pH meter is not available some useful measurements can be made using the antimony electrode (as described in Appendix 2).

Subsequent demonstrations (in which students can play an active part) are concerned mainly with strength of acids (experiment 15.4a) and with titration curves (experiment 15.4b).

Experiment 15.4a
To distinguish strong and weak acids by pH measurement

The teacher will need:

pH meter and electrodes. (Alternatively antimony electrode and $Ag(c)\,|\,Ag^+(aq)$ reference electrode can be used as described in Appendix 2. Either a valve voltmeter or a potentiometer is needed for e.m.f. measurement.*)
1.0M, 0.1M, 0.01M, and 0.001M solutions of acetic acid and hydrochloric acid
Further solutions of acetic acid of molarity 0.5, 0.2, 0.05, and 0.02, to enable a total of eight measurements to be made with this acid

---

*Physics departments may have a Unilab Electrometer – d.c. Amplifier (3.811) which if fitted with a suitable input resistance (16.211) can be calibrated for pH measurements. The apparatus can also be adapted for e.m.f. measurements and for radioactive decay experiments.

Procedure

Measure the pH of the first set of solutions. Wash the electrodes between measurements and start from the most dilute solution to minimize the effect of contamination.

Discuss the values obtained in terms of dilution and possible increased ionization on dilution (see *Students' Book*).

Now measure the pH of the second set of solutions and calculate values of K_a from this. For all but the highest and lowest concentrations, K_a should lie between 1.5 and 2.5×10^{-5}, and at the extremes of the concentration range studied it should not be far outside these limits. If the antimony electrode method is used results may be higher although much depends on the care taken in calibration. An example of this type of calculation for formic acid is given in the *Students' Book*.

The reverse calculation, of pH from K_a, can conveniently be introduced at this stage. The *Students' Book* contains such a problem. This problem can be worked out in a similar way to the calculation of the value of K_a for formic acid from the pH of a formic acid solution, but it involves a quadratic equation. It is worth while to point out to students that an approximation can simplify the equation greatly. This problem, and its solution, are as follows:

Calculate the pH of a 0.001M solution of aminoacetic acid (glycine),

$$NH_2CH_2CO_2H \; (K_a \; = \; 1.7 \times 10^{-10} \; mol \; dm^{-3})$$

$$NH_2CH_2CO_2H(aq) \rightleftharpoons NH_2CH_2CO_2^-(aq) + H^+(aq)$$

$$K_a = \frac{[NH_2CH_2CO_2^-(aq)]_{eqm}[H^+(aq)]_{eqm}}{[NH_2CH_2CO_2H(aq)]_{eqm}}$$

Neglecting the hydrogen ions which arise from ionization of the water, since the concentration of these will be very small compared with the concentration of those from the acid, we can say that

$$[H^+(aq)]_{eqm} = [NH_2CH_2CO_2^-(aq)]_{eqm}$$

and $\quad [NH_2CH_2CO_2H(aq)]_{eqm} = 0.001 - [H^+(aq)]_{eqm}$

$$\therefore \; 1.7 \times 10^{-10} = \frac{[H^+(aq)]_{eqm}^2}{10^{-3} - [H^+(aq)]_{eqm}}$$

At this point, the quadratic can be worked out:

$$[H^+(aq)]_{eqm}^2 + (1.7 \times 10^{-10}[H^+(aq)]_{eqm}) - (1.7 \times 10^{-10} \times 10^{-3}) = 0$$

giving a value for the pH of 6.39.

Alternatively, an approximation can be made. The value of $[H^+(aq)]_{eqm}$ will be about 10^{-5} ($[H^+(aq)]_{eqm}$ for a 10^{-2}M solution of formic acid was about 10^{-3} mol dm^{-3} and K_a for formic acid is very much larger than K_a for aminoacetic acid). Thus we can write:

$$1.7 \times 10^{-10} = \frac{[H^+(aq)]_{eqm}^2}{10^{-3}}$$

$$\therefore pH = 6.38$$

Although accurate results cannot be expected, the pH of the purest available water could be measured at some convenient stage. Discussion of the deviation from the expected value of 7 can be useful.

No treatment of strength of bases is recommended. Every weak base has a conjugate acid, and K_a values can cover all possibilities in this area.

pH changes during titration

This section can be completed by consideration of the changes in pH which take place during titration, a subject which is not dealt with in the *Students' Book*. Two possible ways of doing this are suggested.

1   The *Students' Book* contains titration curves for the four possible systems in Appendix II to this topic. Titrations of the four pairs of acids and bases, running 0.1M base into 25 cm^3 of 0.1M acid, using phenolphthalein and methyl orange as alternative indicators, could be shown to the students and the discussion then based on the titration curves given in the *Students' Book*. The results of the experiments, especially where indeterminate end-points are obtained, are very useful when discussing the titration curves.

2   With good preparation and student assistance, however, the establishment of the experimental basis for a titration curve should not be too time consuming, especially if a pH meter is available. If this is not the case, the antimony electrode and a valve voltmeter (or potentiometer) can be used; time can be saved by preparing the calibration curve beforehand. The titration curves are in an appendix to this topic in the *Students' Book* so that it is less probable that the students will know the results of the experiment before they see it.

Experiment 15.4b
pH changes during titration

The teacher will need:

> pH meter and electrodes (or apparatus for antimony electrode method – see Appendix 2 of this Guide)
> 0.1M hydrochloric acid
> 0.1M acetic acid
> 0.1M sodium hydroxide solution
> 0.1M ammonia solution (keep well stoppered)
> 10 burettes, 50 cm³ $\Big\}$ assuming preliminary titrations done by students – see below
> 10 pipettes, 25 cm³
> 8 titration flasks
> Methyl orange solution
> Phenolphthalein solution
> Small electric stirrer
> Beakers, 100 cm³ (one for each titration curve plotted)

Procedure

1   Titrate 25 cm³ portions of each acid solution with each alkali solution, using as indicator

  a   phenolphthalein

  b   methyl orange

This involves eight titrations in all. These could be done by students whilst the teacher is adjusting the apparatus for pH measurements. The results should be collected together on the blackboard.

2   Pipette 25 cm³ of acid solution into a 100 cm³ beaker and insert electrodes connected to the pH meter. Read the pH of the solution. Add 0.1M alkali from a burette in the approximate quantities shown below, stir well after each addition, and read the pH value. Record the volumes of alkali added and the corresponding pH values. Convenient points in the titration at which to take measurements are

> *Alkali added/cm³* 0, 10, 15, 20, 24, 24.2, 24.4, 24.6, 24.8, 24.9, 25.0, 25.1, 25.2, 25.4, 25.6, 26, 30, 35.

This series assumes that the preliminary titrations, using the appropriate indicator (phenolphthalein for weak acid, methyl orange for weak base) gave a 1:1 ratio for volume of acid:volume of alkali. If this was not the case the above measurement points must be adjusted accordingly.

It is desirable to cover at least two of the possible four systems in this way:

> HCl(aq)/NaOH(aq)

and *either* HCl(aq)/NH$_3$(aq) *or* CH$_3$CO$_2$H(aq)/NaOH(aq) would be suitable choices.

The class can then plot the volume of alkali added (on the horizontal axis) against the pH value (on the vertical axis) for each system investigated.

Discussion

This development does not depend on experiment 15.4b having been seen. The graphs in Appendix II in the *Students' Book* can be used provided that the titrations using indicators have been seen. The calculations could be omitted with the less able students.

In the graphs of volume of alkali added against pH, it is the regions of rapid pH change that are of interest to us. It will be seen that in figure 15.8a in Appendix II of the *Students' Book* (strong acid/strong base) there is a rapid change of pH from about 3 to about 10 due to a very small addition of base solution around the 'end-point'. This range is smaller in figure 15.8b (strong acid/weak base) and covers approximately pH 3–7. This is also the case in figure 15.8c (weak acid/strong base) but here the pH range of rapid change is approximately 7–11. For the weak acid/weak base system there is no marked horizontal portion in the curve and hence, no major region of rapid pH change.

This difference in the pH range at which equal volumes of solution react completely (25 cm^3 of each in this case) focuses attention on what exactly we mean by the end-point of a titration. In the examples under discussion, the solutions are all of the same molarity, and equal volumes should react together. The end-point therefore means the *equivalence-point* when the quantities of substance specified in the equation have reacted together. If this point can be made to correspond with a colour change in a suitable indicator we have a means of detecting the end-point easily and simply. Experience shows that choice of the correct indicator for use in acid/base titrations depends on the type of system used.

1   For strong acids with strong bases almost any indicator can be used.

2   For strong acids with weak bases the choice is restricted; methyl red or methyl orange are commonly used.

3   For weak acids with strong bases the choice is again restricted but indicators different from those in (2) must be used; phenolphthalein is a common choice.

4   For weak acids with weak bases it is very difficult to find a suitable indicator.

The suitability of indicators for different purposes was established by trial and error long before the idea of pH was introduced. From such information we might confidently guess that the colour change with methyl red or methyl orange occurs somewhere in the pH range 3–7 (figure 15.8b), and for phenolphthalein somewhere in the pH range 7–11 (figure 15.8c).

In cases where the use of indicators is unsuitable or impossible, a pH meter provides a simple method of following an acid–base titration.

From the titration curves it is obvious that the equivalence-point in a titration does not invariably correspond with the production of a neutral solution, which would be defined as one with pH 7. In the strong acid/strong base (figure 15.8a) system a solution very near to this pH value is obtained. With a strong acid and a weak base (figure 15.8b) the mixture is acid (low pH), and with a weak acid and a strong base (figure 15.8c) it is alkaline (high pH). To understand the reasons for these differences we must look more closely at conditions when reacting quantities of acid and base are present in the three examples.

For the strong acid/strong base example, the reaction can be written

$$Na^+(aq) + OH^-(aq) + H^+(aq) + Cl^-(aq) \rightarrow H_2O(l) + Na^+(aq) + Cl^-(aq)$$

Here the $Na^+(aq)$ and $Cl^-(aq)$ ions have no very pronounced acidic or basic properties and the equilibrium established is

$$H^+(aq) + OH^-(aq) \rightleftharpoons H_2O(l)$$

for which, as we have seen, the pH is 7 when $[H^+(aq)]_{eqm} = [OH^-(aq)]_{eqm}$.

With the strong acid/weak base system we can represent the reaction as

$$NH_3(aq) + H^+(aq) + Cl^-(aq) \rightarrow NH_4^+(aq) + Cl^-(aq)$$

Again the $Cl^-(aq)$ ion has no marked acidic or basic properties but the $NH_4^+(aq)$ ion is a not insignificant acid, which dissociates to $NH_3(aq)$ and $H^+(aq)$

$$NH_4^+(aq) \rightleftharpoons NH_3(aq) + H^+(aq)$$

For this equilibrium $K_a = 6 \times 10^{-10}$ and from this we can calculate the pH of the solution obtained at equivalence-point in the titration represented in figure 15.8b. Here 25 cm³ 0.1M HCl(aq) has reacted with 25 cm³ 0.1M NH₃(aq). The result will be 50 cm³ of 0.05M NH₄Cl(aq). Since K_a for the equilibrium shown above is small we shall not be seriously in error if we put $[NH_4^+(aq)]_{eqm} = 0.05$ mol dm⁻³. Also, if we neglect the effect of the $OH^-(aq)$ ions from the ionization of the water (this will have a very small influence on the $H^+(aq)$ ion concentration), we can assume that

$$[NH_3(aq)]_{eqm} = [H^+(aq)]_{eqm}$$

Thus

$$K_a = \frac{[NH_3(aq)]_{eqm}[H^+(aq)]_{eqm}}{[NH_4^+(aq)]_{eqm}}$$

$$6 \times 10^{-10} = \frac{[H^+(aq)]^2_{eqm}}{0.05}$$

$$\therefore [H^+(aq)]_{eqm} = \sqrt{(6 \times 10^{-10} \times 0.05)}$$

$$= 5.5 \times 10^{-6}$$

$$-\lg [H^+(aq)]_{eqm} = -(\overline{6}.74)$$

$$= -(-6+0.74)$$

$$= 5.26$$

$$\therefore \text{ pH of solution at equivalence-point } = 5.26$$

This agrees with figure 15.8b where pH 5.25 is at about the mid-point of the horizontal section of the titration curve.

For the weak acid/strong base in figure 15.8c the equation can be written

$$Na^+(aq)+OH^-(aq)+CH_3CO_2H(aq) \rightarrow Na^+(aq)+CH_3CO_2^-(aq)+H_2O(l)$$

Neglecting $Na^+(aq)$ this reduces to

$$OH^-(aq)+CH_3CO_2H(aq) \rightleftharpoons CH_3CO_2^-(aq)+H_2O(l)$$

The acetate ion is a comparatively strong base and the equilibrium position is not entirely over to the right. A calculation similar to that given above shows that the pH of 0.05M sodium acetate solution is 8.75, which agrees well with the mid-point of the horizontal portion of the graph in figure 15.8c.

For the ammonia/acetic acid titration in figure 15.8d the pH at equivalence-point will be the average of the two values given above, i.e. $\frac{5.25+8.75}{2} = 7.00$.

The fact that this coincides with the neutral point is accidental. If formic acid is titrated with ammonia solution the pH at equivalence-point is 6.75. The value varies with the K_a values for the acid systems involved in the equilibria established at the equivalence-point.

It would be useful at this stage to add universal indicator solution to solutions of the salts obtained in the acid–base reactions which have been studied (NaCl, NH_4Cl, CH_3CO_2Na, $CH_3CO_2NH_4$) and check that the above argument is reasonable.

Changes in pH value at the end-point of a titration

The origin of the rapid changes in pH at the end-point of the strong acid/strong base titration (figure 15.8a) can be seen from a simple calculation of the pH value one drop (about 0.05 cm^3) before the equivalence-point, and one drop after the equivalence-point.

For one drop before the equivalence-point there will be 25.00 cm^3 0.1M HCl(aq) present and 24.95 cm^3 0.1M NaOH(aq), i.e. an excess of 0.05 cm^3 0.1M HCl(aq) in 49.95 cm^3 (near enough to 50 cm^3) total solution

$$H^+(aq) \text{ ions per 50 cm}^3 = \frac{0.05 \times 0.1}{1000} \text{ mol}$$

$$\therefore [H^+(aq)] = \frac{0.05 \times 0.1 \times 1000}{1000 \times 50} \text{ mol dm}^{-3}$$

$$= 10^{-4} \text{ mol dm}^{-3}$$

$$\therefore \text{ pH value } = 4$$

One drop after equivalence-point we have 0.05 cm^3 0.1M NaOH(aq) in excess

$$OH^-(aq) \text{ ions per 50 cm}^3 = \frac{0.05 \times 0.1}{1000} \text{ mol}$$

$$\therefore [OH^-(aq)] = \frac{0.05 \times 0.1 \times 1000}{1000 \times 50} = 10^{-4} \text{ mol dm}^{-3}$$

but $$[H^+(aq)][OH^-(aq)] = K_w = 10^{-14}$$

$$\therefore [H^+(aq)] = \frac{10^{-14}}{10^{-4}} = 10^{-10}$$

$$\therefore \text{ pH value } = 10$$

Thus there is a change in pH value of 6 units during the addition of two drops of 0.1M NaOH(aq) at the end-point of this titration.

For the titration of a strong acid with a weak base (figure 15.8b) the corresponding rapid change in pH extends over the range 3–7 only, and for a weak acid and a strong base (figure 15.8c) over the range 7–10. For a weak acid/weak base system there is no region where the pH change is of this magnitude.

Suggestions for homework

Calculations on pH values of weak acid solutions of different molarities.

Summary

At the end of this section students should be expected to be able to discuss acid–base theory on a qualitative basis, and to identify species which are acting as acids and bases in a given system. They should know what is meant by pH and K_a, be able to discuss the pH changes that take place during acid–base reactions, and relate these to the K_a values of the acids concerned.

15.5 Buffer solutions and indicators

Objectives

1 To introduce the idea of a buffer solution.

2 To develop an equation to use for calculations involving pH of buffer solutions.

3 To discuss the simple theory of indicators and to measure K_a for an indicator and for a weak acid.

Timing

About four periods.

Suggested treatment

For this treatment OP transparencies numbers 54 and 121 will be useful.

The subjects of buffer solutions and indicators are dealt with in some detail in the *Students' Book* and this might be read by students beforehand. Little more needs to be added here. If a pH meter is available the effect of addition of acid and alkali to a buffer solution and to a non-buffer solution of approximately the same pH could be shown. Also a comparison of calculated and measured pH values for a given buffer solution would be of interest. In experiment 15.5 students determine K_a for an indicator and for a weak acid.

Experiment 15.5

To measure K_a for (a) an indicator and (b) a weak acid

Each student or pair of students will need:

> 18 test-tubes, 125×16 mm (or equivalent size specimen tubes)
> 1 rack to hold the test-tubes in pairs one behind the other so that the colour can be seen through each pair of tubes
> 1 teat pipette to deliver approximately 0.5 cm^3
> 1 measuring cylinder (25 cm^3)

access to:

> Bromophenol blue solution (0.1 g dissolved in 20 cm^3 ethanol and then made up to 100 cm^3 with water)
> Concentrated hydrochloric acid
> Approximately 4M sodium hydroxide solution

and communal burettes containing:

> 0.02M formic acid (made up approximately from solution from suppliers and then standardized)
>
> 0.02M sodium formate (1.36 g HCO_2Na per dm^3)
>
> 0.02M benzoic acid (2.44 g $C_6H_5CO_2H$ per dm^3. Make up in warm water and allow to cool)
>
> 0.02M sodium benzoate (2.88 g $C_6H_5CO_2Na$ per dm^3)

Procedure

Students make up solution X which contains bromophenol blue indicator in the HIn form by adding one drop of concentrated hydrochloric acid to 5 cm^3 of indicator solution. They also make up a solution in which the indicator is in the In^- form by adding one drop of 4M sodium hydroxide solution to 5 cm^3 of indicator solution.

They then set up nine pairs of tubes so that the colour *when looking through both tubes* is that due to the indicator with the ratio of the $[HIn]:[In^-]$ form, 1:9, 2:8, 3:7, and so on.

Students then find which pair of tubes matches most closely the indicator colour produced by mixing 5 cm^3 of 0.02M sodium formate, 5 cm^3 of 0.02M formic acid, and 10 drops of bromophenol blue solution. The answers to the questions at the end of the instructions in the *Students' Book* should be as follows.

   1   The pair of tubes with the $[HIn]/[In^-]$ ratio = 6/4 will probably match the colour of the solution best. This is assumed in the following answers.

   2   The pH of the solution is given by

$$pH = -\lg K_a - \lg \frac{[acid]_{eqm}}{[base]_{eqm}}$$

Assuming all the formate ions come from the sodium formate:

$$pH = -\lg (2 \times 10^{-4}) - \lg \frac{0.01}{0.01}$$

$$= 3.7$$

   3   The ratio of $[HIn]_{eqm}/[In^-]_{eqm}$ at this pH is 6/4.

   4   K_a for bromophenol blue is given by

$$pH = -\lg K_a - \lg \frac{[HIn(aq)]_{eqm}}{[In^-(aq)]_{eqm}}$$

$$3.7 = -\lg K_a - \lg \frac{6}{4}$$

$$\lg K_a = -3.88$$

$$K_a = 1.3 \times 10^{-4}$$

Students then measure K_a for benzoic acid by finding which pair of tubes matches most closely the indicator colour produced by mixing 5 cm^3 of 0·02M sodium benzoate, 5 cm^3 of 0.02M benzoic acid, and 10 drops of bromophenol blue. Answers to the questions given after this experiment should be as follows.

    1 The pair of tubes with the $[HIn]_{eqm}/[In^-]_{eqm}$ ratio = 3/7 will probably match the colour of this solution best. This is assumed in the following answers.

    2 The $[HIn]_{eqm}/[In^-]_{eqm}$ ratio is 0.429.

    3 The pH of the mixture is given by:

$$pH = -\lg K_a - \lg \frac{[HIn(aq)]_{eqm}}{[In^-(aq)]_{eqm}}$$

$$= -\lg (1.3 \times 10^{-4}) - \lg 0.429$$

$$= 4.26$$

  4 K_a for benzoic acid is given by

$$pH = -\lg K_a - \lg \frac{[acid]_{eqm}}{[base]_{eqm}}$$

$$4.26 = -\lg K_a - \lg \frac{0.01}{0.01}$$

$$\lg K_a = -4.26$$

$$K_a = 5.5 \times 10^{-5}$$

The accuracy of these results depends on the accuracy with which the solutions have been made up, and also on the size of the gap between the sets of tubes. It would be instructive for students to calculate the value of K_a in each case using the $[HIn]_{eqm}/[In^-]_{eqm}$ ratio for the pair of tubes *next* to the pair that they used.

Formulae for indicators
Students will probably be curious about these. A few examples are given in figure 15.5.

Methyl orange

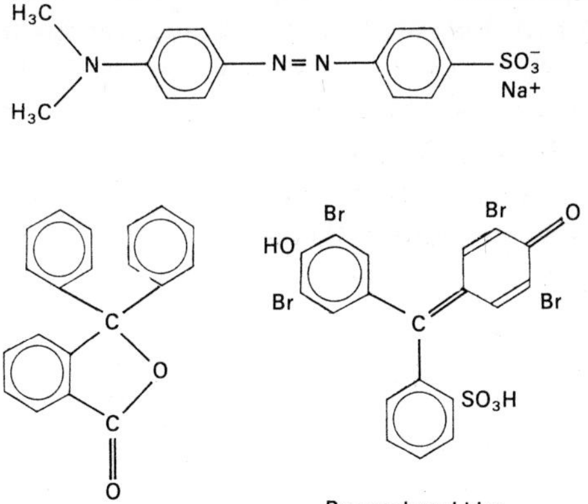

Bromophenol blue

Phenolphthalein

Figure 15.5

Enquiries about litmus should be referred to an article by H. G. Andrew on the subject (*School Science Review* (1963), 153, 338).

Suggestions for homework

Calculating pH values for buffer solutions of given composition and vice versa.
Looking up uses of buffer solutions.
Choosing indicators for given acid–base titrations.
Reading the section entitled 'Acid–base chemistry in the human body', which is included as background reading at the end of this topic in the *Students' Book*.

Summary

At the end of this section students should be able to:

1 Explain the action of a buffer system.

2 Calculate the pH of buffer solutions when composition is known and vice versa.

3 Understand the elementary theory of indicators.

Answers to problems in the *Students' Book*

(A suggested mark allocation is given in brackets after each answer.)

1

i   Fe(s) reductant, $Cu^{2+}(aq)$ oxidant   (1)

ii   Al(s) reductant, $H^{+}(aq)$ oxidant   (1)

iii   Zn(s) reductant, $Pb^{2+}(aq)$ oxidant   (1)

iv $Sn^{2+}(aq)$ reductant, $Fe^{3+}(aq)$ oxidant   (1)

                                                                                                Total (4)

2

i   Metals   (1)

   All form positive ions in aqueous solution   (2)

ii $D(s)+2C^{+}(aq) \rightarrow D^{2+}(aq)+2C(s)$ (2)

iii   a $A(s) \rightarrow A^{2+}(aq)+2e^{-}$

          $B^{2+}(aq)+2e^{-} \rightarrow B(s)$ (2)

   b $A(s) \rightarrow A^{2+}(aq)+2e^{-}$

          $2C^{+}(aq)+2e^{-} \rightarrow 2C(s)$ (2)

   c $B(s) \rightarrow B^{2+}(aq)+2e^{-}$

          $2C^{+}(aq)+2e^{-} \rightarrow 2C(s)$ (2)

   d $D(s) \rightarrow D^{2+}(aq)+2e^{-}$

          $B^{2+}(aq)+2e^{-} \rightarrow B(s)$ (2)

iv   A, D, B, C   (2)

   A reduces $B^{2+}(aq)$ and $C^{+}(aq)$

   D does not reduce $A^{2+}(aq)$

   D reduces $B^{2+}(aq)$ and

   B reduces $C^{+}(aq)$ (5)

                                                                                                Total (20)

3

i H_2 positive, Fe negative   (1)

   $E^{\ominus} = -0.44$ volt   (2)

ii   Ni negative, H_2 positive   (1)

   $E^{\ominus} = +0.23$ volt   (2)

iii   Zn negative, Ni positive   (1)

   $E^{\ominus} = 0.76-0.23 = +0.53$ volt   (2)

iv   Al negative, Cr positive   (1)

   $E = 1.66-0.74 = +0.92$ volt   (2)

                                                                                                Total (12)

4   Standard electrode potential $= 0.34-0.62$

                                                     $= -0.28$ volt   (2)

5   Standard electrode potential $= 1.61-0.76$

                                                     $= +0.85$ volt   (2)

6

i $Al(s) \rightarrow Al^{3+}(aq) + 3e^-$
$Sn^{2+}(aq) + 2e^- \rightarrow Sn(s)$
$2Al(s) + 3Sn^{2+}(aq) \rightarrow 3Sn(s) + 2Al^{3+}(aq)$ (3)

ii $Pb(s) \rightarrow Pb^{2+}(aq) + 2e^-$
$Ag^+(aq) + e^- \rightarrow Ag(s)$
$Pb(s) + 2Ag^+(aq) \rightarrow 2Ag(s) + Pb^{2+}(aq)$ (3)

iii $Mg(s) \rightarrow Mg^{2+}(aq) + 2e^-$
$2H^+(aq) + 2e^- \rightarrow H_2(g)$
$Mg(s) + 2H^+(aq) \rightarrow Mg^{2+}(aq) + H_2(g)$ (3)

Total (9)

7

i $Ag^+(aq)$ $Cu^{2+}(aq)$ $Pb^{2+}(aq)$ $Cr^{3+}(aq)$ (2)

ii $Fe^{3+}(aq)$ $Sn^{2+}(aq)$ $Zn^{2+}(aq)$ $Mg^{2+}(aq)$ (2)

Total (4)

8

i   (5)

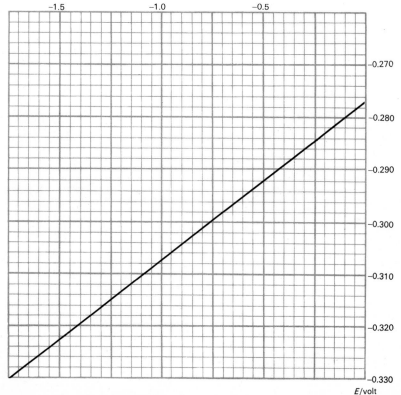

lg [ion]

E/volt

ii -0.277 volt   (2)
     $E = E^{\ominus}$ when $\lg [\text{ion}] = 0$ (3)
iii Slope of graph $= 0.03$ (2)
     z (charge on ion) $= 2$ (2)
     As M is a metal, charge on ion $= M^{2+}$ (1)
                                                                        Total (15)

9

i   From the Nernst equation
     $\lg [Ag^+(aq)] = -3.2$
     $\quad [Ag^+(aq)] = 7 \times 10^{-4}$ mol dm^{-3}
     But marks should be given for correct reading from the actual
     graph. (*Note*. Students must use the graph of *electrode potential*
     against $\lg [Ag^+(aq)]$ on page 76 of the *Students' Book*.)   (2)
ii $K_{sp} = [Ag^+(aq)]_{eqm}[BrO_3^-(aq)]_{eqm}$ (2)
     10 cm^3 of the 0.1M KBrO$_3$(aq) precipitate
     $\therefore$ 40 cm^3 remain in a total volume of 60 cm^3 (2)

     $\therefore [BrO_3^-]$ in final solution $= \dfrac{40}{60} \times 10^{-1}$ mol dm^{-3} (2)

     $\therefore K_{sp} = [Ag^+(aq)] \times \dfrac{1}{15}$ mol^2 dm^{-6} (2)
                                                                        Total (10)

10 $K_{sp} = [Ag^+(aq)]_{eqm}[CNS^-(aq)]_{eqm} = 2 \times 10^{-12}$ mol^2 dm^{-6} (2)
        $[Ag^+(aq)]_{eqm} = \sqrt{2 \times 10^{-12}}$ mol dm^{-3} (2)
        1 mole AgCNS $= 166$ g   (2)
        $\therefore$ One cubic decimetre of solution contains
        $166 \times \sqrt{2 \times 10^{-12}}$ g of AgCNS   (2)
        $= 2.4 \times 10^{-4}$ g   (2)
                                                                        Total (10)

11 1 mole CaC$_2$O$_4$ $= 128$ g   (1)
        $[CaC_2O_4] = \dfrac{0.068}{128}$ mol dm^{-3} (2)

        $[Ca^{2+}(aq)]_{eqm} = [C_2O_4^{2-}(aq)]_{eqm} = \dfrac{0.068}{128}$ (2)
        $K_{sp} = [Ca^{2+}(aq)]_{eqm}[C_2O_4^{2-}(aq)]_{eqm}$ (2)
        $\quad = \left(\dfrac{0.068}{128}\right)^2$ (1)
        $\quad = 2.8 \times 10^{-7}$ mol^2 dm^{-6} (2)
                                                                        Total (10)

12 $K_{sp} = [Fe^{3+}(aq)]_{eqm} \times [OH^-(aq)]_{eqm}^3 = 8 \times 10^{-40}$ (2)

$[OH^-(aq)]_{eqm} = 3[Fe^{3+}(aq)]_{eqm}$ (2)

$\therefore K_{sp} = 27[Fe^{3+}(aq)]_{eqm}^4$ (2)

$[Fe(OH)_3(aq)] = [Fe^{3+}(aq)]_{eqm} = \sqrt[4]{\dfrac{8}{27} \times 10^{-40}}$ mol dm^{-3} (3)

1 mole $Fe(OH)_3 = 107$ g (2)

Solubility $Fe(OH)_3 = 107 \times \sqrt[4]{\dfrac{8}{27} \times 10^{-40}}$ (2)

$\qquad = 7.9 \times 10^{-9}$ g dm^{-3} (2)

Total (15)

13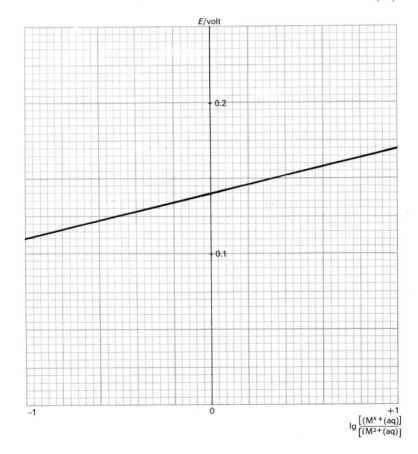

E/volt

$\lg \dfrac{[(M^{x+}(aq)]}{[(M^{2+}(aq)]}$

Graph $\dfrac{[M^{x+}]}{[M^{2+}]}$ against E $\hspace{4cm}$ (5)

$E^{\ominus} = 0.14$ volt $\hspace{5cm}$ (2)

$E^{\ominus} = E$ when $\lg \dfrac{[M^{x+}]}{[M^{2+}]} = 0$ $\hspace{3cm}$ (3)

Slope of graph $= 0.03$ $\hspace{4.5cm}$ (2)

$\therefore z =$ number of electrons transferred when oxidized form
$\hspace{3cm}$ changes to reduced form $= 2$ $\hspace{2cm}$ (2)

$\therefore x = 4$ $\hspace{6.5cm}$ (1)

$\hfill$ *Total* (15)

14

i   Acid–base $\hspace{7cm}$ (1)

$$C_2H_5CO_2H(l) + H_2O(l) \rightleftharpoons C_2H_5CO_2^-(aq) + H_3O^+(aq)$$
$\hspace{1cm}$ (acid$_1$) $\hspace{1.5cm}$ (base$_2$) $\hspace{1.5cm}$ (base$_1$) $\hspace{1.5cm}$ (acid$_2$) $\hspace{1cm}$ (4)

ii Redox $\hspace{7.5cm}$ (1)

$$Pb(s) + 2Ag^+(aq) \rightleftharpoons Pb^{2+}(aq) + 2Ag(s)$$
(reductant$_1$) $\hspace{0.5cm}$ (oxidant$_2$) $\hspace{0.5cm}$ (oxidant$_1$) $\hspace{0.5cm}$ (reductant$_2$) $\hspace{0.5cm}$ (4)

iii Redox $\hspace{7.5cm}$ (1)

$$Mg(s) + 2H^+(aq) \rightleftharpoons Mg^{2+}(aq) + H_2(g)$$
(reductant$_1$) $\hspace{0.5cm}$ (oxidant$_2$) $\hspace{0.5cm}$ (oxidant$_1$) $\hspace{0.5cm}$ (reductant$_2$) $\hspace{0.5cm}$ (4)

iv Acid–base $\hspace{7cm}$ (1)

$$H_3O^+(aq) + OH^-(aq) \rightleftharpoons 2H_2O(l)$$
$\hspace{1cm}$ (acid$_1$) $\hspace{1.5cm}$ (base$_2$) $\hspace{1cm}$ (base$_1$ and acid$_2$) $\hspace{1cm}$ (4)

$\hfill$ *Total* (20)

15

i $[H^+(aq)]_{eqm} = 2 \times 10^{-1}$ mol dm^{-3} $\hspace{3cm}$ (1)

$\hspace{0.5cm}$ pH $= -\lg 2 \times 10^{-1}$ $\hspace{4cm}$ (1)

$\hspace{0.5cm}$ pH $= 0.7$ $\hspace{5.5cm}$ (1)

$\hfill$ *Total* (3)

(In answers to this, and other parts of question 15 it is likely that many students will not show the steps by which they reached the pH value. Full marks may be given for the correct answer alone, on the understanding that no marks can be given for an incorrect answer alone. The marking scheme is structured so that credit can be given for partially correct answers in which the working *is* shown.)

ii $[OH^-(aq)]_{eqm} = 2 \times 10^{-1}$ mol dm^{-3} $\hspace{3cm}$ (1)

$\hspace{0.5cm}$ $[H^+(aq)]_{eqm} = 10^{-14} \div (2 \times 10^{-1})$ $\hspace{2.5cm}$ (1)

$\hspace{0.5cm}$ pH $= -\lg 5 \times 10^{-14}$ $\hspace{3.5cm}$ (1)

$\hspace{0.5cm}$ pH $= 13.30$ $\hspace{5cm}$ (1)

$\hfill$ *Total* (4)

iii $[H^+(aq)]_{eqm} = 1.25 \times 10^{-1} \text{ mol dm}^{-3}$ (1)
     $pH = -\lg 1.25 \times 10^{-1}$ (1)
     $pH = 0.90$ (1)
                                                                      Total (3)

iv Excess of acid $\equiv 50 \text{ cm}^3$ 0.1M in a total volume of 100 cm^3 (1)
     Concentration of acid $= 0.05M$ (1)
     $[H^+(aq)]_{eqm} = 5 \times 10^{-2} \text{ mol dm}^{-3}$ (1)
     $pH = -\lg 5 \times 10^{-2}$ (1)
     $pH = 1.30$ (1)
                                                                      Total (5)

v $CH_2BrCO_2H(aq) \rightleftharpoons CH_2BrCO_2^-(aq) + H^+(aq)$ (1)
     $$K_a = \frac{[CH_2BrCO_2^-(aq)]_{eqm}[H^+(aq)]_{eqm}}{[CH_2BrCO_2H(aq)]_{eqm}}$$ (1)
     $\quad = 1.35 \times 10^{-3}$
     Let $[H^+(aq)]_{eqm} = x$
     Then $\dfrac{x^2}{0.1-x} = 1.35 \times 10^{-3}$ (3)
     If x is small $0.1 - x \approx 0.1$
     $\therefore x^2 = 1.35 \times 10^{-3} \times 10^{-1}$
     $\quad = 1.35 \times 10^{-4}$ (1)
     $\therefore x = 1.16 \times 10^{-2} \text{ mol dm}^{-3}$ (1)
     $\therefore pH = -\lg 1.16 \times 10^{-2} = 1.94$ (1)
                                                                      Total (8)

16 For
     $HCO_2H(aq) \rightleftharpoons HCO_2^-(aq) + H^+(aq)$
     $K_a = 2 \times 10^{-4}$
     Neglecting the hydrogen ions which arise from ionization of the water,
     $\qquad [HCO_2^-(aq)]_{eqm} = [H^+(aq)]_{eqm}$
     $\qquad [HCO_2H(aq)]_{eqm} = 0.01 - [HCO_2^-(aq)]_{eqm}$
     $\therefore \dfrac{[HCO_2^-(aq)]_{eqm}^2}{0.01 - [HCO_2^-(aq)]_{eqm}} = 2 \times 10^{-4}$ (1)
     Ignoring $[HCO_2^-(aq)]_{eqm}$ with respect to 0.01:
     $\dfrac{[HCO_2^-(aq)]_{eqm}^2}{0.01} = 2 \times 10^{-4}$ (1)
     $[HCO_2^-(aq)]_{eqm} = 1.4 \times 10^{-3} \text{ mol dm}^{-3}$ (1)
                                                                      Total (3)

17 For

$$HA(aq) \rightleftharpoons H^+(aq) + A^-(aq)$$

$$K_a = \frac{[H^+(aq)]_{eqm}[A^-(aq)]_{eqm}}{[HA(aq)]_{eqm}}$$

and $\quad [A^-(aq)]_{eqm} = [H^+(aq)]_{eqm} = 1.3 \times 10^{-3} \text{ mol dm}^{-3}$

$\therefore [HA(aq)]_{eqm} = [0.1-(1.3 \times 10^{-3})] \text{ mol dm}^{-3}$

$$\therefore K_a = \frac{(1.3 \times 10^{-3})^2}{0.1-(1.3 \times 10^{-3})} \tag{1}$$

Ignoring 1.3×10^{-3} with respect to 0.1,

$$K_a = \frac{(1.3 \times 10^{-3})^2}{0.1} \tag{1}$$

$$= 1.7 \times 10^{-5} \tag{1}$$

Total (3)

18 $pH = -\lg [H^+(aq)]_{eqm} = 5.1$

$\therefore [H^+(aq)]_{eqm} = 7.94 \times 10^{-6} \tag{1}$

For $\quad HA(aq) \rightleftharpoons H^+(aq) + A^-(aq)$

$[H^+(aq)]_{eqm} = [A^-(aq)]_{eqm} = 7.94 \times 10^{-6} \text{ mol dm}^{-3}$

$[HA(aq)]_{eqm} = [0.1-(7.94 \times 10^{-6})] \text{ mol dm}^{-3}$

$$\therefore K_a = \frac{[H^+(aq)]_{eqm}[A^-(aq)]_{eqm}}{[HA(aq)]_{eqm}}$$

$$= \frac{(7.94 \times 10^{-6})^2}{0.1-(7.94 \times 10^{-6})} \tag{1}$$

Ignoring 7.94×10^{-6} with respect to 0.1

$$K_a = \frac{(7.94 \times 10^{-6})^2}{0.1} \tag{1}$$

$$= 6.3 \times 10^{-10} \tag{1}$$

Total (4)

19 For the equilibrium

$$C_6H_5NH_3^+(aq) \rightleftharpoons H^+(aq) + C_6H_5NH_2(aq)$$

$[C_6H_5NH_2(aq)]_{eqm} = [H^+(aq)]_{eqm}$

$[C_6H_5NH_3^+(aq)]_{eqm} = 0.001 - [H^+(aq)]_{eqm}$

$$\therefore \frac{[H^+(aq)]_{eqm}^2}{0.001-[H^+(aq)]_{eqm}} = 3 \times 10^{-5} \tag{1}$$

Ignoring $[H^+(aq)]_{eqm}$ with respect to 0.001

$[H^+(aq)]_{eqm} = 1.73 \times 10^{-4} \text{ mol dm}^{-3} \tag{1}$

$pH = -\lg [H^+(aq)]_{eqm} = -(-3.8)$

$$= 3.8 \tag{1}$$

Total (3)

Topic 16
Some d-block elements

Objectives

To study some aspects of d-block metal chemistry, selecting various elements from the first row to illustrate

1   variable oxidation number
2   formation of complex ions
3   catalytic properties.

Content

16.1 Introduction to d-block elements. Variable oxidation number.
16.2 Complex ions.
16.3 d-Block elements in catalysis.

Timing

Two weeks.

16.1   Introduction to d-block elements
Variable oxidation number

Objectives

1   To introduce the d-block elements.

2   To outline reasons for their variable oxidation numbers.

3   To make students familiar with the main oxidation numbers of some d-block elements.

4   To investigate some of the reactions of the compounds of vanadium experimentally, to observe changes in oxidation number of this element.

Timing

Six periods.

Suggested treatment

For the introduction to this section, the teacher should have available:

Samples of d-block elements and some of their alloys.

Samples of a selection of compounds of the d-block elements in which these elements have as many different oxidation numbers as possible.

Models of body-centred and face-centred cubic structures (for details of their construction see Appendix 3).

OP transparencies numbers 37, 79, 80, 114–116.

It will also be helpful to have an energy level display board (previously used in Topic 4) and an oxidation number display boards (see Topic 5). Details of the construction of these boards are given in Appendix 3.

To introduce the study of d-block elements, the following points could be covered.

1   This block of ten elements is part of the 'd-block'. There is some controversy over whether zinc should be included. Although it appears as the last element in the break from the 2, 3, 3 pattern in the ionization energy/atomic number curve, as Zn^{2+} has a full 3d level, much of the chemistry of zinc is more similar to s-block elements than to other d-block elements. Similarly scandium only exists with an oxidation number of three or zero and is more similar to aluminium than to other d-block elements. A reasonably satisfactory definition of a d-block element is *one which forms some compounds in which there is an incomplete inner subshell of d electrons*. This definition includes the elements from titanium to copper but not scandium or zinc. When the students know what is meant by a d-block element, the properties of these elements can be discussed in more detail.

2   All these elements have many similar properties.

a   All are metals. X-ray diffraction techniques have given us an idea of the structure of pure metal crystal lattices and alloy lattices. As the d-block elements have atoms with small diameters, the metals have small atomic volumes and so have high densities and are hard materials. As they have similar atomic diameters, they can form alloys with each other (see table 16.1a in the *Students' Book*). The structures and uses of three metals, titanium, chromium, and copper are discussed in the Background reading at the end of section 16.1 in the *Students' Book*. Reading this could form part of a homework.

b   All have coloured ions, that is, their ions absorb light in the visible part of the spectrum. The electronic energy levels of ions of other elements are separated by distances such that the absorption of energy to promote electrons is in the ultraviolet part of the spectrum. But some of the energy levels in d-block elements are such that visible light can bring about excitation of electrons. This can happen in two ways.

Just as the second energy level is split into two sub-shells, 2s and 2p, in atoms other than hydrogen, so the d level is split into two when a complex ion is formed. The difference in energy between these two levels is normally such that visible light can excite an electron from the lower to the higher level. Thus complex ions involving d-block elements, such as $Cu(H_2O)_4^{2+}$ or $Fe(CN)_6^{3-}$, absorb visible light.

Visible light is also absorbed when electrons are transferred from a ligand to a metal to form a new structure. This accounts for the colour of the permanganate ion.

c   All form many complex ions. These are discussed in section 16.2.

d   All exhibit catalytic activity. This is discussed in section 16.3.

e   All exhibit variable oxidation number. A discussion of this property is set out in the *Students' Book*, and it is suggested that students read this for part of a homework; it can then be discussed in class.

f   Iron, cobalt, and nickel are ferromagnetic.

Observation of the samples of the elements, their compounds, and the Periodic Table will bring out these points.

3   In Topic 4 it was found that the 2, 3, 3 grouping of first ionization energies was broken after calcium, and that the ten transitional elements produced this break. Also that a new energy level, the d level, belonging to the n = 3 quantum level became occupied after calcium. This level comes just above the 4s level and below the 4p level, as indicated in figure 16.1a.

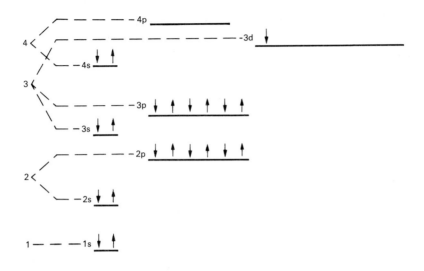

Figure 16.1a
Electronic structure of scandium.

As the atomic number increases from 21 to 30 each element has one more electron in the 3d level than the previous elements, until the d level is full (10 electrons); then gallium has a full 3d level and one electron in the 4p level.

The teacher may care to note that the electrons do not fill the 3d level in a straightforward manner, as in some cases an arrangement with only one electron in the 4s level and an extra one in the 3d level may be more stable. In fact the 3d level is below the 4s from scandium onwards but the two levels are very close in energy. The teacher should use as much of this information as he sees fit. All students should realize that the 3d and 4s levels are very close and that therefore the most stable arrangement of electrons may not necessarily have two electrons in the 4s level.

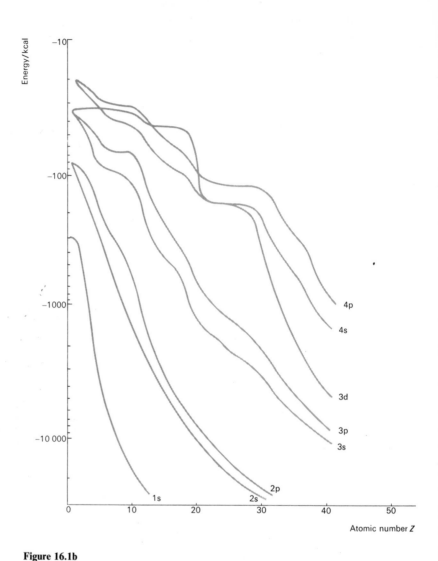

Figure 16.1b

Change of energy level with atomic number. (1 kcal = 4.2 kJ)
After Harvey, K. B. and Porter, G. B. (1963) Introduction to physical inorganic chemistry, *Addison Wesley*

After these general points have been discussed, the students can investigate changes in oxidation number further by means of experiment 16.1a.

Experiment 16.1a
Investigation of the oxidation numbers of vanadium

Each student will need:

Test-tubes
Copper powder (copper(II) oxide *reduced just before use*)
Zinc dust

access to solutions having the following approximate concentrations:

0.1M ammonium vanadate (about 10 g solid dissolved in 160 cm^3 water + 80 cm^3 of concentrated sulphuric acid followed by dilution to 1 dm^3)
0.1M iron(II) ammonium sulphate
0.1M iron(III) ammonium sulphate
0.1M potassium iodide
0.05M iodine in 0.2M potassium iodide solution
0.1M potassium bromide
0.1M sodium thiosulphate
Saturated sulphur dioxide solution
Solutions in bottles having teat pipettes as stoppers will be most convenient.

Students can make their own solutions of V(III), V(IV), and V(II) from V(V), but it may be more convenient for the teacher to provide these after experiments 1 to 5 have been done. They should be made as in 3 to 5, the sulphur dioxide boiled out in 3, and the other two solutions, after warming, left on the solids. The students should be told how these solutions have been made so that they can explain such observations as the formation of a precipitate in 4.

Procedure
The experiment is described in the *Students' Book*. It will first be necessary for the students to find how $E^\ominus$ values can be used to predict the possible courses of reactions. It is suggested that they read the appropriate section in the *Students' Book* for homework so that it can then be discussed in class before the experimental work is started.

Substances to be mixed	Prediction	Experimental result
1 V(V) + Fe(II)	→ Fe(III) + V(IV)	√
2 V(V) + I(−I)	→ I(0) + V(IV)	√
3 V(V) + S(IV)	→ S(VI) + V(III)	→ S(V) + V(IV)
4 V(V) + Cu(0)	→ Cu(II) + V(III)	√
5 V(V) + Zn(0)	→ Zn(II) + V(II)	√ (eventually)

In (2) iodine is produced and students should remove it to see the colour of the vanadium ion. Sodium thiosulphate solution should be used for this purpose.

In (3) the predicted reaction on thermodynamic grounds is to S(VI) and V(III) but the reaction S(V) to S(VI) is slow and so for kinetic reasons the reaction stops at S(V). V(IV) is therefore formed.

In (4) if copper powder containing copper(II) oxide is used, some copper(I) will be formed and a precipitate will appear.

The reaction mixture, especially in (3) to (5), may require warming.

Students should also observe the series of colours: yellow, V(V)(aq), through green to blue, V(IV)(aq), to green, V(III)(aq), to violet, V(II)(aq), in (5). When they have completed this section, this series of colours can be demonstrated by boiling V(V) with granulated zinc, when the reaction is slower.

	Substances to be mixed	Prediction	Experimental result
6	V(IV) + V(II)	V(III)	√
7	V(V) + V(III)	V(IV)	√
8	V(IV) + V(III)	no reaction	√
9	V(IV) + Fe(III)	no reaction	√
10	V(III) + Fe(III)	→ V(IV) + Fe(II)	√
11	V(IV) + I(−I)	no reaction	√
12	V(V) + Br(−I)	no reaction	√
13	V(III) + I(O)	→ V(IV) + I(−I)	√
14	V(IV) + I(O)	no reaction	√

Teachers who have time might care to spend two to three periods looking more closely at the chemistry of one d-block element in more detail. Experiment 16.1b on the oxidation numbers of manganese might be appropriate for this.

Experiment 16.1b
Investigation of the oxidation numbers of manganese

Each student will need:

Card, rubber bands, and specimen tubes to make a demonstration display chart of manganese compounds (see 'Supporting material', Topic 5) made of stiff cardboard on which small specimen tubes can be held with rubber bands
Test-tubes

access to:

    2M sulphuric acid
    2M sodium hydroxide
    0.01M potassium permanganate
    Manganese(II) sulphate
    Manganese(IV) oxide
    Potassium hydroxide pellets
    Concentrated sulphuric acid

Suggested treatment

A demonstration display chart of manganese compounds (see Topic 5) should be set up with samples of solid potassium permanganate, manganese(IV) oxide, manganese(II) sulphate pentahydrate and, if possible, metallic manganese (as shown in figure 16.1c). Solutions should also be on the chart. Students can then make their own miniature charts and as more oxidation states are prepared, they can be added to the charts.

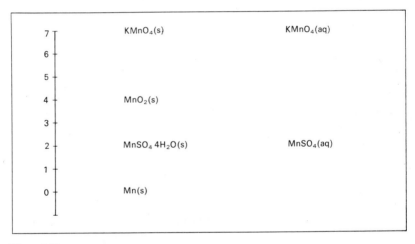

Figure 16.1c
An oxidation number display chart for manganese.

Manganese(VI)

Mixing Mn(VII) and Mn(IV) might give Mn(VI). $E^{\ominus}$ values indicate that this reaction is not feasible under standard conditions of acidity and as the $E^{\ominus}$ values are far apart, change of conditions will not reverse the reaction. Further, increase of acidity will favour the reverse reaction.

Under alkaline conditions, $E^{\ominus}$ values still do not favour the reaction but as they are very close, change of conditions might reverse the reaction. Increase of alkalinity should tip the balance in favour of the reaction we require.

Students find that the only mixture that reacts is that which is alkaline; after filtering, a green solution of Mn(VI) remains. Addition of acid brings about the expected reaction:

$$Mn(VI) \rightarrow Mn(IV) + Mn(VII)$$

Manganese(III)

Mixing Mn(II) and Mn(IV) is a possibility here. The $E^\ominus$ values under acid conditions indicate that the reaction is not feasible. A lower pH might tip the balance but the $E^\ominus$ difference is quite high. The reaction should occur under alkaline conditions; increase of pH will not affect the reaction. As the reaction under these conditions is between crystalline $Mn(OH)_2$ and MnO_2, it is likely to be very slow. Furthermore, students will observe that on standing $Mn(OH)_2$ is oxidized by atmospheric oxygen, not to $Mn(OH)_3$ but to MnO_2.

Another possibility which involves species one of which is soluble in water under all conditions of pH, is to mix Mn(II) and Mn(VII). The $E^\ominus$ values indicate that this reaction is just feasible under acid conditions and decrease of pH will favour the reaction. Mn(VII) is normally reduced to Mn(II) in a redox reaction.

Students will obtain red Mn(III) by this method and, on dilution with water, raising the pH will cause disproportionation into Mn(IV) and Mn(II) as might be expected from the equation.

Manganese(V)

The $E^\ominus$ values indicate that preparation from the two most likely oxidation numbers available, Mn(IV) and Mn(VI), under acid or alkaline conditions is not feasible and that pH will not affect the reaction. Students should be shown the preparation and can then try to explain what is occurring.

The simplest way is to add a crystal or two of potassium permanganate to a concentrated solution of potassium hydroxide (about 12M; 30 pellets solid potassium hydroxide in 2 cm^3 water). This should be done by the teacher wearing safety spectacles, and the hazards of fused or concentrated alkali discussed.

The solution will slowly become blue and close observation will show that a gas is being given off. Although the rate of evolution is such that it would be difficult to identify this gas, it is oxygen.

Assuming that Mn(VII) has been reduced to Mn(V), what has been oxidized to what? The oxygen is the clue: students may require some help in deciding what is happening. Oxygen is being oxidized from $(-\text{II})$ to (O)

$$2OH^- \rightarrow H_2O + \tfrac{1}{2}O_2 + 2e^-; \quad E^{\ominus} = -0.4 \text{ V}$$

Most oxidizing agents, which thermodynamically can bring this oxidation about, normally do not do so as it is controlled kinetically.

Students can then repeat the experiment at a lower pH and obtain green Mn(VI). Having the $E^{\ominus}$ value above, they should be able to deduce that the reducing agent is again $O(-\text{II})$.

Homework suggestions
Writing balanced ionic equations for the reactions that have taken place.

In view of the fact that natural waters can be acid, neutral, or alkaline, that rocks contain Mn(II), and that the air contains oxygen, students might be interested in looking up the forms in which manganese occurs in nature.

Summary
Students should know what are the most important oxidation numbers of elements of the first row of the d-block, and be able to use a redox potential chart to predict *possible* reactions.

16.2 Complex ions

Objectives
1 To show that complex ion formation is similar to acid-base equilibria and that the size and charge of ions affects the stability of the ion.
2 To study some complexes formed by copper(II).
3 To investigate the number of ligands complexing with a metal cation in solution by measuring the colour intensity of the solution as the proportion of ligand to metal cation is increased.

Timing
Four periods.

Suggested treatment
For this treatment, OP transparencies numbers 51 and 53 will be useful.

The *Students' Book* contains a discussion of how complex ion formation is similar to acid–base equilibrium but with cations other than H^+. A survey of two factors which affect the stability of complex ions, the size of the central ion, and its charge, then follows, and finally the stability constant is introduced as a measure of the stability of a complex ion.

Experiment 16.2a

Investigation of some complexes

Each student will need:

Test-tubes
Teat pipettes

access to solutions having the following approximate concentrations:

Concentrated hydrochloric acid
Concentrated ammonia solution
0.5M copper(II) sulphate
0.2M EDTA (disodium salt)
0.1M sodium 2-hydroxybenzoate (salicylate)
0.1M 1,2-dihydroxybenzene (catechol) in 0.5M sodium hydroxide (*this solution should be freshly prepared*)

Procedure

The experiments are described in the *Students' Book*. It is *most important* that students think about the questions at the end of each section of practical work and answer them, perhaps in discussion, before going on to the next section of the practical work. Without this, practical work will have little or no meaning.

Blue $Cu(H_2O)_4^{2+}$ ions are present in a solution of copper(II) and addition of concentrated hydrochloric acid produces some yellow $CuCl_4^{2-}$ ions giving a green solution. One can think of a mixture of the two ions or, more accurately, a gradual substitution of Cl^- for H_2O round the copper(II) ion. Addition of a large quantity of water produces a blue colour again as the equilibrium is disturbed. Addition of ammonia gives dark blue $Cu(NH_3)_4^{2+}$. These reactions are to be expected from the stability constants, which also indicate that addition of EDTA solution to the $Cu(NH_3)_4^{2+}$ solution will give the paler blue of the $Cu(II)$–EDTA complex.

The stability constants also predict that when sodium 2-hydroxybenzoate is added to the solution of deep blue $Cu(NH_3)_4^{2+}$, the green $Cu(II)$-2-hydroxybenzoate complex will be formed. With EDTA solution, this gives the pale blue $Cu(II)$–EDTA complex, and this with 1,2-dihydroxybenzene gives the green $Cu(II)$-1,2-dihydroxybenzene complex.

Experiment 16.2b

Investigation of the stoichiometry of the Ni(II) – EDTA complex ion

Each student or pair of students will need:

1 colorimeter with 2 tubes
11 test-tubes

access to communal burettes containing :

0.05M nickel(II) sulphate
0.05M EDTA (disodium salt)
(These concentrations may require modification for the particular colorimeters used)

Procedure

Students make up solutions containing Ni(II) and EDTA in molar proportions $0:10$, $1:9$, $2:8$, and so on, and then determine $\lg \frac{I_0}{I}$, which is proportional to the molarity of the complex, for each mixture. From a graph of these readings, the mixture containing the highest concentration of complex can be found. The proportions in this mixture are those of the complex. For the Ni(II)-EDTA ion they will be found to be $1:1$.

At this point it would be profitable to spend part of a homework reading in the *Students' Book* about some aspects of the stereochemistry of complexes and about some uses of complexing agents. These can then be discussed in class.

Summary

Students should understand how complex formation occurs, how stability constants can be used, and how the stoichiometry of some complex ions can be determined.

Note. For the teacher's benefit, it is worth noting that the equilibrium constant for complexing with EDTA is different from that for complexing with Cl^-. The equation for the disodium salt of EDTA and Cu^{2+} is:

$$(CH_2)_2N_2(CH_2CO_2H)_2(CH_2CO_2)_2^{2-}(aq) + Cu^{2+}(aq) \rightarrow$$
$$(CH_2)_2N_2(CH_2CO_2)_4Cu^{2-}(aq) + 2H^+(aq)$$

so the equilibrium constant involves $[H^+]$. The equilibrium constant for complexing with Cl^- does not involve $[H^+]$.

16.3   d-Block elements in catalysis

Objectives

1   To discuss the function of a catalyst.
2   To investigate experimentally the use of ions of d-block elements to catalyse a reaction.
3   To look at some of the uses of d-block elements in industrial catalysis.

Timing

Four periods.

Suggested treatment

Students will already have come across the idea of catalysis. In this section they investigate whether the reaction between persulphate and iodide ions is catalysed by ions of elements of the first period of d-block elements. They can then try to account for their results. Redox reactions should be to the forefront of their minds.

Experiment 16.3

A study of the catalytic action of d-block ions on the reaction between iodide and persulphate ions

Each pair of students will need:

> 1 stopclock, or sight of large clock with seconds hand
> 1 100 cm^3 or 150 cm^3 conical flask
> 1 boiling tube
> Distilled water

access to communal burettes containing:

> Saturated potassium persulphate solution
> 0.2M potassium iodide
> 0.01M sodium thiosulphate
> (This solution is best made up shortly before use, either from the solid, or by dilution of 0.1M solution)
> 0.2 per cent starch solution

access to solutions containing the following ions (all concentrations are approximate; solutions are best supplied in bottles with integral teat pipettes):

> 0.1M Cr(III)
> 0.1M Cr(VI)
> 0.1M Ni(II)
> 0.1M Co(II)
> 0.1M Cu(II)
> 0.1M Fe(II)
> 0.1M Fe(III)
> 0.1M Mn(II)
> 0.1M Mn(VII)
> Also a solid Cu(I) salt

Procedure

The details are in the *Students' Book*. Discussion of the errors involved in the results is essential for students to decide whether catalysis has occurred. They will find that Mn(VII), Fe(III), Fe(II), and Cu(II) catalyse the reaction. Discussion should then bring out the possibility that catalysis is by means of alternate oxidation and reduction of the catalyst ion and this suggests two questions:

1 Does the higher oxidation number oxidize I($-$I) to I(O)?

2 Does persulphate oxidize the lower oxidation number to the higher?

Redox potentials can be used to check whether these reactions are possible and then a practical check can be made to find out whether the kinetics are such that these reactions actually occur. Results are given in table 16.3.

	Cr		Mn		Fe		Co		Ni		Cu	
	VI	**III**	**VII**	**II**	**III**	**II**	**III**	**II**	**IV**	**II**	**II**	**I**
Does it catalyse?	×	×	√	×	√	√	?	×	?	×	√	?
Does higher number oxidize I⁻?												
Redox potential	√		√		√		√		√		√*	
Practically	√		√		√		√‡		√‡		√	
Does $S_2O_8^{2-}$ oxidize lower number?												
Redox potential		√		√		√		√		√		√
Practically		×		√†		√		×		×		√

* $e^- + Cu^{2+}(aq) + I^-(aq) \rightarrow CuI(s)$.
† very slowly.
‡ Co(III) and Ni(IV) can be electrochemically generated and both oxidize I⁻ to I^2

Table 16.3

Immediately we see that only those metals which have ions with the higher oxidation number which oxidize I⁻, and ions with the lower number which reduce $S_2O_8^{2-}$, catalyse the reaction. Of course other oxidation numbers may be involved and catalyse the reaction – this may explain the observed results with manganese.

It should be made clear to students that this is only a *possible* mechanism – other mechanisms may explain the facts equally well or even better.

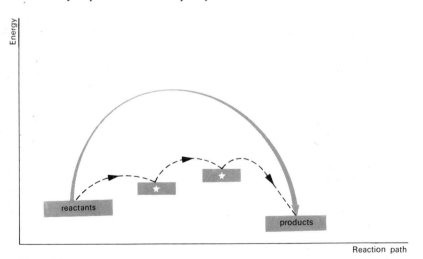

Figure 16.3
The effect of a catalyst on a reaction.
——— reaction path without catalyst, activation energy E_1
– – – reaction path with catalyst, activation energy E_2
  * intermediate products of the catalysed reaction

It is worth pointing out to students that the action of a catalyst can be explained by assuming that it enables the reaction to proceed by *a new path involving a lower activation energy* and that it is regenerated during the reaction. This may be explained by a diagram similar to that in figure 16.3.

By this time students should have read the background material on trace elements in the *Students' Book* and this should have been discussed in class.

Summary
Students should realize the extent of the use of catalysts, and how possible theories of their mechanism can be tested.

Topic 17
Equilibrium and free energy

Introduction

The main purpose of this topic is to link up and extend various parts of the chemical experience students already have. In Topic 12 (Equilibria: gaseous and ionic) and in Topic 15 (Equilibria: redox and acid–base systems) they have learned the importance of the equilibrium constant as a means of describing the direction and extent of chemical change in equilibrium systems. In Topic 7 (Energy changes and bonding) they have learned how to forecast on an approximate basis, using enthalpies of formation, whether or not a given chemical compound is likely to be stable at room temperature. In Topic 12 they have also been introduced to the connection between standard enthalpies and the way in which the values of equilibrium constants alter with temperature change. In Topic 14 (Reaction rates) they have considered the kinetic aspects of the stabilities of substances. In Topic 16 (Some d-block elements) they have again tackled the problem of forecasting the direction and extent of chemical reactions, this time of ions in solution, in terms of standard redox potentials ($E^{\ominus}$ values).

In this topic, therefore, the aim is to draw together these ideas, to relate energy changes more precisely to equilibrium constants, and to discuss in more detail those questions centred upon the problem of determining the direction of chemical changes, including the stabilities of substances and the feasibility of chemical reactions. The topic thus has an important integrative function, as the following Venn diagram shows (figure 17.A).

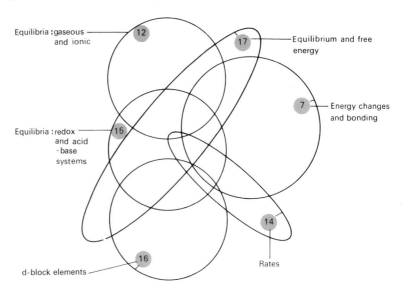

Figure 17.A
Venn diagram illustrating the connections between Topics 7, 12, 14, 15, 16, and 17.

In terms of the themes running through the course (see Appendix 5, page 311), Topic 17 is also a focal point. All seven major themes are to be found here, including some important applications of chemistry.

Objectives
When they have reached the end of this topic students should:

1 Understand that the fundamental tendency in any chemical process is the attainment of equilibrium.

2 Understand the relationship between the equilibrium constant and $\Delta H^\ominus$ and be able to find equilibrium constants for the same (gas) reaction at different temperatures.

3 Understand the relationship between the equilibrium constant and $\Delta G^\ominus$ and be able to find equilibrium constants indirectly for both reactions between gases and reactions between ions in solution.

4 Be able to use tables of the standard free energy of formation of compounds as a means of calculating equilibrium constants indirectly.

Content
17.1 The direction of a chemical change.

17.2 How to find equilibrium constants, for any particular (gas) reaction, at different temperatures.

17.3 How to find equilibrium constants (at a particular temperature) indirectly.

17.4 Free energy in chemistry.

Timing
Between two and three weeks will be required for this topic.

17.1 **The direction of a chemical change**

Objectives

The aim of this section is to make clear to the students the following three main points:

1   The fundamental tendency in chemical systems is the tendency to attain equilibrium.

2   This tendency applies also to reactions which 'go to completion' or 'do not occur'. In these cases K has very large or very small values.

3   In practice, kinetic barriers may frequently prevent the attainment of equilibrium.

Timing

One period will be enough for this introductory section.

Introduction

In this section students are reminded of a number of points, touched on in earlier topics, including the value of equilibrium constants as a means of expressing the direction and extent of a chemical reaction (Topic 12), and the fact that kinetic factors may frequently prevent the attainment of equilibrium in chemical systems.

Suggested treatment

The idea of the direction of a chemical change seems at first sight a clear and unambiguous one. We are familiar with many chemical reactions which 'go virtually to completion' or 'do not occur to any appreciable extent'. The words 'virtually' and 'appreciable' indicate that the cases we are thinking of, for instance the combustibility of heated carbon or the refractory nature of magnesium oxide when heated, are extremes.

It is however possible to cite many examples of chemical reactions which do not 'go virtually to completion' nor 'fail to occur to any appreciable extent'. The *Students' Book* quotes the equilibrium

$$N_2O_4(g) \rightleftharpoons 2NO_2(g)$$

with which students are already familiar from Topic 12 and reminds students that here:

1   The fundamental tendency of the system is to attain a state of dynamic equilibrium.

2   One convenient way of expressing the position of equilibrium is by means of the equilibrium constant (K_c or K_p).

Glancing back to the examples at the beginning of this section we see that we can extend to them this idea that systems tend to attain equilibrium by noting that in such cases the position of equilibrium is 'far to the right' (K very large) or 'far to the left' (K very small). Quoting a few calculated values of K for such reactions (10^{69} for the reaction to form carbon dioxide from its elements for example) shows that we are justified in such special cases in speaking of reactions going virtually to completion or failing to occur to any appreciable extent.

Kinetic barriers to chemical reaction

It is important that students should be reminded once again of a point discussed in Topic 14 (Reaction rates), that although there is a tendency for chemical systems to attain equilibrium, there are often kinetic factors which prevent this. The treatment in the *Students' Book* points out that in order to do any conversion of reactants to products, the conditions must be such that a substantial proportion of the reacting species are supplied with energy of a magnitude comparable with that of the *activation energy* for the process. This can be done

a   by heating the reactants

or b   by finding a catalyst which will effect the conversion by an alternative route for which the activation energy is lower and comparable with the energies of the reactants at ordinary temperatures

or   by a combination of both (a) and (b).

If the kinetic barrier is not overcome, mixtures of substances can coexist for indefinitely long periods in a state far removed from equilibrium. It is important to keep this constantly in mind, especially in this topic. Our very existence depends upon it.

Summary

At the end of this section students should appreciate that:

1   The fundamental tendency in chemical systems is the tendency to attain equilibrium.

2   This applies also to reactions which 'go to completion' (K very large) or 'do not occur' (K very small).

3   In practice, kinetic barriers may frequently prevent the attainment of equilibrium.

17.2   **How to find equilibrium constants, for any particular (gas) reaction, at different temperatures**

Objective

To enable students to understand the relationship between K_p and $\Delta H^\ominus$ for a (gas) reaction, and, given the value of the equilibrium constant, K_p, at one particular temperature, and the value of $\Delta H^\ominus_{298}$ for the reaction, to calculate values for K_p at other temperatures.

Timing

Two or three periods will be needed for this section.

Introduction

In Topic 12 (Equilibria: gaseous and ionic) students have already noted the general way in which a change of temperature affects the change in the value of the equilibrium constant for a gas reaction and how this is related to the sign of $\Delta H^\ominus$ for the reaction. This section takes the discussion a stage further and shows how T, K_p, and $\Delta H^\ominus$ are in fact related. This will enable students to calculate for themselves values of K_p for gas reactions at temperatures at which the equilibrium constants have not been directly determined.

Suggested treatment

This section could be started by reminding students of the generalization they met in Topic 12 when discussing the effect of temperature on the equilibrium constant for a gas reaction, namely, '. . . an *increase in temperature* of an equilibrium system results in an *increase in the value of the equilibrium constant* if the reaction involved is *endothermic* ($\Delta H^\ominus$ positive), and a *decrease in the value of the equilibrium constant* if the reaction is *exothermic* ($\Delta H^\ominus$ negative)'.

At that stage it was merely noted that the sign of $\Delta H^\ominus$ affected the way in which K_p changed with change of temperature. We now wish to establish not only how the *sign* of $\Delta H^\ominus$ affects this, but also how the *magnitude* of the $\Delta H^\ominus$ value is involved.

The students could be asked to look at some values of K_p at various temperatures and to attempt to discover the relation between K_p, T, and $\Delta H^\ominus$. The method will depend very much on the ability and experience of the students and on the personal choice of the teacher. An able class could simply be given the values of T, K_p, and $\Delta H^\ominus$ for various reactions and left to work out the relationship for themselves. A more average class may find the suggested procedure in the *Students' Book* more useful. For this procedure, a number of values of equilibrium constants for some gaseous reactions at different temperatures are given in table 1 in the appendix to this topic in the *Students' Book*. It would probably be best to divide the data of table 1 between the students in groups so that all the data are worked on and the findings subsequently pooled.

The *Students' Book* suggests that students first try plotting K_p against T and where they run into difficulties over the K_p scale they then plot lg K_p against T. They may recognize that the resulting curves are sections of hyperbolae, or at least that an inverse relationship seems to be implied between lg K_p and T. The next move therefore is to plot lg K_p against $\dfrac{1}{T}$.

Students are sometimes worried about taking logarithms of physical quantities, which have units, since it is usually understood that one can take the logarithm only of a pure number. Suggesting that students plot the logarithm of K_p, which frequently has units (atmospheres, raised to an appropriate power, depending on the form of the equilibrium equation) may precipitate the difficulty.

One way of dealing with this is to consider the expression lg K_p to be a shorthand way of writing lg $\left(\dfrac{K_p}{1}\right)$, that is, the *ratio* of the value of K_p at present being considered to the value unity for K_p. A ratio is a pure number and its logarithm can therefore be taken with impunity. For example, for the equilibrium

$$N_2O_4(g) \rightleftharpoons 2NO_2(g)$$

$K_p = \dfrac{p_{NO_2}^2}{p_{N_2O_4}}$, and therefore has units of atmospheres. 'lg K_p' should be taken to mean lg $\left(\dfrac{K_p \text{ atm}}{1 \text{ atm}}\right)$ so that the *numerical* value of K_p is the value of the ratio, the logarithm of which is taken.

It is next necessary to quote a result that cannot be derived at this level, namely

$$\text{lg } K_p \approx \text{constant} - \frac{\Delta H^{\ominus}}{2.3\,R}\left(\frac{1}{T}\right)$$

Using their graphs of lg K_p against $\dfrac{1}{T}$ students can, however, verify that the gradients of their graphs are in general agreement with the values obtained from $-\dfrac{\Delta H^{\ominus}}{2.3\,R}$

It should be pointed out that the relationship

$$\lg K_p \approx \text{constant} - \frac{\Delta H^{\ominus}}{2.3\,R}\left(\frac{1}{T}\right)$$

is an approximation, but so long as the temperature range over which the relationship is assumed to hold is not too wide, it is a good approximation. The reason why it is an approximation is as follows.

The exact relation between K_p, T, and $\Delta H^{\ominus}$ is

$$\frac{\text{d}\ln K_p}{\text{d}T} = \frac{\Delta H^{\ominus}}{RT^2}$$

which can be rewritten*

$$\frac{\text{d}\ln K_p}{\text{d}\left(\frac{1}{T}\right)} = -\frac{\Delta H^{\ominus}}{R} \qquad (1)$$

This implies that, if K_p is determined at various temperatures, and a plot is made of $\ln K_p$ against $\frac{1}{T}$, the slope of the plot will be $-\frac{\Delta H^{\ominus}}{R}$. The *sign* of $\Delta H^{\ominus}$ therefore determines the *direction* of the slope, and the *magnitude* of $\Delta H^{\ominus}$ the *steepness* of the slope. This will be true whatever is assumed about the variation of $\Delta H^{\ominus}$ with temperature. If $\Delta H^{\ominus}$ is effectively constant with change of temperature, the plot will be a straight line. If the temperature range is wide enough, however, the variation of $\Delta H^{\ominus}$ with temperature will lead to a slight curvature, but the gradient of the curve at any given temperature can still be determined.

Students could be asked how closely they find their plots of $\lg K_p$ against $\frac{1}{T}$ approximate to straight lines and whether any curvature is detectable.

---

* Put $x = \frac{1}{T}$ $\therefore T = \frac{1}{x}$ that is $\frac{\text{d}T}{\text{d}x} = -\frac{1}{x^2} = -T^2$.

$$\frac{\text{d}\ln K_p}{\text{d}x} = \frac{\text{d}\ln K_p}{\text{d}T} \times \frac{\text{d}T}{\text{d}x}$$

$$= \frac{\Delta H^{\ominus}}{RT^2} \times -T^2$$

$$\therefore \frac{\text{d}\ln K_p}{\text{d}\left(\frac{1}{T}\right)} = \frac{-\Delta H^{\ominus}}{R}$$

When it comes to integrating equation (1) it is necessary to make some assumption about the variation of $\Delta H^{\ominus}$ with temperature. The simplest assumption is that $\Delta H^{\ominus}$ is independent of temperature, in which case the integrated form of the equation is

$$\ln K_p = \text{constant} - \frac{\Delta H^{\ominus}}{R}\left(\frac{1}{T}\right)$$

It is because, very strictly speaking, $\Delta H^{\ominus}$ is not independent of temperature that this equation is not exact. It is, however, a good approximation and the closeness of the approximation depends on how good our assumption of the temperature-independence of $\Delta H^{\ominus}$ is. Students can see how good an assumption this is by:

1   Looking for departures from linearity in their plots of $\lg K_p$ against $\frac{1}{T}$.

2   Looking at some experimental values of $\Delta H_T^{\ominus}$ over various wide temperature ranges. Values for a number of reactions are given in table 2 in the appendix to this topic in the *Students' Book*.

A way in which students could do this is suggested in the *Students' Book*.

Calculating values of K_p at various temperatures

The way in which this part of the section is taught will, again, depend on the ability of the students. The *Students' Book* approaches the problem by first considering an example in which a straightforward linear interpolation from a graph of $\lg K_p$ against $\frac{1}{T}$ is used to calculate K_p at a value of T for which it has not been determined experimentally.

A value of 2.5 for $\lg K_p$ and hence of 316 for K_p at 800 K can be obtained from a graph such as that shown in figure 17.2.

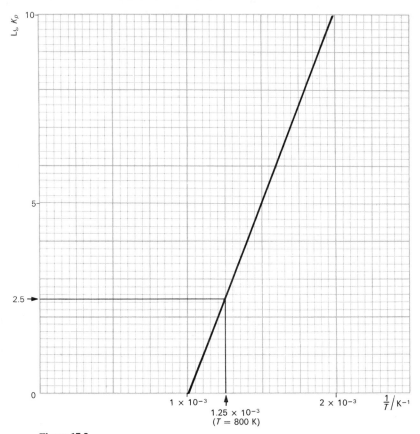

Figure 17.2

Graph of $\lg K_p$ against $\dfrac{1}{T}$ for the equilibrium $2SO_2(g) + O_2(g) \rightleftharpoons 2SO_3(g)$.

When $T = 800 \text{ K}$, $\dfrac{1}{T} = 1.25 \times 10^{-3} \text{ K}^{-1}$ and, from this graph $\lg K_p = 2.5$, whence $K_p(800 \text{ K}) = 316$.

A second example shows how, given one value of K_p at a particular temperature, the value of $\Delta H_{298}^{\ominus}$, and the relation

$$\lg K_p = \text{constant} - \frac{\Delta H^{\ominus}}{2.3\,R}\left(\frac{1}{T}\right),$$

the value of K_p at any other temperature (within a reasonable range) can be calculated, as follows.

Substitute values of 2.5×10^{10} atm^{-1}, -197 kJ mol^{-1}, 8.31 J K^{-1} mol^{-1}, and 500 K, for K_p, ΔH , R, and T:

$$\lg(2.5 \times 10^{10}) = \text{constant} - \frac{-197}{2.3 \times 8.31 \times 10^{-3}} \times \frac{1}{500}$$

$$10.40 = \text{constant} + 20.6$$

$$\therefore -10.2 = \text{constant}$$
(to 3 significant figures)

Using this value of the constant for this reaction and now inserting $T = 800$ K, and other values as before in the same equation

$$\lg K_p = -10.2 - \frac{-197}{2.3 \times 8.31 \times 10^{-3}} \times \frac{1}{800}$$

$$= -10.2 + 12.9$$

$$= +2.7$$

$$\therefore K_p = 501 \text{ atm}^{-1}$$

This answer should be compared with that obtained by the graphical method. The discrepancy in the values of $\lg K_p$ (2.7 compared with 2.5) is about eight per cent. The discrepancy between the resulting values of K_p is, however, about fifty per cent, the difference between $10^{2.7}$ and $10^{2.5}$. Moreover, a change by one kilojoule per mole in $\Delta H^{\ominus}$ (-196 instead of -197 kJ mol^{-1}) would give a value of $\lg K_p$ of 2.6, and of K_p itself of 400 atm^{-1}. Students should realize that the exact value of K_p will depend rather sensitively on small differences such as these, but that calculations of this kind can reliably be used to estimate the order of magnitude of K_p (is it of the order of hundreds, thousands, millions, or what?)

Finally, a general form of example 2 is derived, namely

$$\lg \frac{K_2}{K_1} = \frac{\Delta H^{\ominus}}{2.3\,R} \left(\frac{1}{T_1} - \frac{1}{T_2} \right)$$

Students can then compare this expression with the original generalization quoted from Topic 12 (Equilibria: gaseous and ionic) at the beginning of this section, in order to note:
1   That the mathematical expression agrees with the verbal formula.
2   That the mathematical expression not only tells us the general trend of K_p values with temperature change for exo- and endothermic processes; it also enables us to calculate values for equilibrium constants at various temperatures using not only the information given by the *sign* of $\Delta H^{\ominus}$ but also the information given by its *magnitude*.

Supporting homework

Students could usefully tackle some of the problems (see the end of this topic) on the material of this section.

Summary

At the end of this section, students should understand the relationship between K_p and $\Delta H^{\ominus}$, and be able to find equilibrium constants for any particular (gas) reaction at different temperatures.

17.3   How to find equilibrium constants (at a particular temperature) indirectly

Objectives

1   To introduce to students a *definition* of the standard free energy change, $\Delta G^{\ominus} = -2.3\, RT \lg K$.

2   To recall the use of $E^{\ominus}$ values to predict the likely direction of change of a chemical reaction and to show how equilibrium constants can be calculated from $E^{\ominus}$ values using the Nernst equation.

3   To picture $\Delta G^{\ominus}$ as $-zFE^{\ominus}$, the limiting work obtainable from an electrochemical cell.

4   To get students to appreciate that there are ways of obtaining values of $\Delta G^{\ominus}$ independent of the determination of equilibrium constants, so that these $\Delta G^{\ominus}$ values can be used to calculate K values indirectly.

5   (Optional.) To discuss entropy and chemical reactions.

Timing

Between eight and twelve periods will be needed for this section, depending on the students' previous experience, and on whether the teacher wishes to pursue the subject of entropy (see below).

Introduction

Having, in the previous section, considered ways of calculating equilibrium constants at temperatures at which they have not been measured directly, we now continue the discussion of the indirect determination of equilibrium constants, this time at a fixed temperature (for convenience, 298 K). In this way we shall extend the range of methods available to us for determining equilibrium constants and hence for forecasting the likely direction of chemical change.

Summary of suggested teaching sequence

Aim: to extend our under-→ Define $\Delta G^{\ominus}$ as →Explore the logical conse-
standing of indirect methods $-2.3\ RT\ \lg K$ quences of this. $\Delta G^{\ominus}$ large
of determining K.   and negative when K is very
                                                     large, and so on.

Use of $E^{\ominus}$ values to forecast $\Delta G^{\ominus}$ and $\Delta H^{\ominus}$. A glance
the direction of chemical   back to Topic 7 where we
change (Topic 16).   assumed that when $-\Delta H^{\ominus}$
                                                     is large, K is large, when $\Delta H^{\ominus}$
                                                     is large K is small, and so on.
                                                     (Optional: entropy)

More precise statement of   Proper role of $\Delta H^{\ominus}$ is as
relation between $E^{\ominus}$ and K_c,   indicated in section 17.2.
as

$$E^{\ominus} = \frac{2.3\ RT}{zF}\lg K_c$$

Comparison of indirect   Comparison of→Other methods of determin-
(from e.m.f.s) and direct   these two equa- ing $\Delta G^{\ominus}$ lead to still further
(titrimetric) methods of find- tions leads to ways of finding K.
ing value of K_c.   relation
                              $$\Delta G^{\ominus} = -zFE^{\ominus}$$

                                                     17.4 Examples of use of $\Delta G^{\ominus}$
                                                     values to calculate K.

—➤— main sequence

----➤--- other connections

Suggested treatment

For this treatment OP transparencies numbers 96 and 97 will be useful.

After students have been reminded that this topic is about ways of finding K values indirectly, they can be presented with the relationship between $\Delta G^{\ominus}$ and K:

$$\Delta G^{\ominus} = -2.3 \, RT \lg K$$

Such a definition of $\Delta G^{\ominus}$ needs some justification and it is useful to make the following points at the outset:

1 Since RT has units of energy (for example, kJ mol^{-1}) and $\lg K$ is a pure number, $\Delta G^{\ominus}$ also has units of energy.

2 $\Delta G^{\ominus}$ is a temperature-dependent quantity since T is in this equation as well as $\lg K$. $\Delta G^{\ominus}$ values must therefore be quoted for some particular temperature. A useful standard temperature at which many values are quoted is 298 K (25 °C), and such standard free energy changes at this temperature are written $\Delta G^{\ominus}_{298}$.

3 The 2.3 enters this equation because the relation between $\Delta G^{\ominus}$ and K, as derived 'from advanced theory' is $\Delta G^{\ominus} = -RT \ln K_p$.

4 Students may ask why the relationship is a logarithmic one. This is difficult to answer at this stage, but one can point out that K values are multiplicative whereas energy values are additive, so that a logarithmic relation seems reasonable. Examples of the multiplicative property of equilibrium constants which could be quoted are

a $N_2O_4(g) \underset{}{\overset{K_1}{\rightleftharpoons}} 2NO_2(g)$

$K_3 \updownarrow \qquad\qquad \updownarrow K_2$

$2NO(g) \; + \; O_2(g)$

for which $K_3 = K_1 \times K_2$

b   successive ionization of polybasic acids such as oxalic acid

$H_2C_2O_4(aq) \underset{K_1}{\rightleftharpoons} HC_2O_4^-(aq) + H^+(aq)$

$HC_2O_4^-(aq) \underset{K_2}{\rightleftharpoons} C_2O_4^{2-}(aq) + H^+(aq)$

overall:

$H_2C_2O_4(aq) \underset{K_3}{\rightleftharpoons} C_2O_4^{2-}(aq) + 2H^+(aq)$

for which $K_3 = K_1 \times K_2$

It is important to remember that it is the *standard* free energy change, $\Delta G^{\ominus}$, that is related to the equilibrium constant of a chemical reaction in this way and not any other free energy change ΔG. This point is best made with the students later (section 17.4). In the meantime it should be referred to as 'the standard free energy change' or 'delta G standard'.

Having 'defined' $\Delta G^{\ominus}$ in this way, we can now examine how the value of $\Delta G^{\ominus}$ by definition depends on the value of K. Figure 17.3a shows a plot of $\Delta G^{\ominus}$ against lg K at 298 K. (A graph of $\Delta G^{\ominus}$ against K itself on a linear scale is difficult to draw because of the vast scale required to accommodate a meaningful range of K values.) Several points are clear from an inspection of figure 17.3a.

1   For very large values of K (greater than about 10^{10}), $\Delta G^{\ominus}_{298}$ is negative and has values of -60 kJ mol^{-1} or less. This corresponds to a situation in which the reaction has gone 'virtually to completion' (see section 17.1).

2   For very small values of K (less than about 10^{-10}) $\Delta G^{\ominus}_{298}$ is positive and has values of 60 kJ mol^{-1} or more. This corresponds to a situation in which the reaction has 'failed to occur'.

3   In the intermediate range (10^{-10} to 10^{10}) of values of K, which includes all those we usually refer to as equilibrium reactions, $\Delta G^{\ominus}_{298}$ has values in the range $+60$ kJ mol^{-1} to -60 kJ mol^{-1}. For equilibria in which the position of equilibrium is such that reactants predominate, $\Delta G^{\ominus}_{298}$ has small positive values. For equilibria in which the position of equilibrium is such that products predominate, $\Delta G^{\ominus}_{298}$ has small negative values. We may summarize all this as shown in table 17.3a.

Standard free energy change $\Delta G^{\ominus}_{298}$/kJ mol^{-1}	Equilibrium constant K_{298}	Extent of reaction
Less than -60	10^{10} and larger	Reaction 'complete'
Between -60 and 0	10^{10} to 1	Products predominate in an 'equilibrium'
Between 0 and $+60$	1 to 10^{-10}	Reactants predominate in an 'equilibrium'
Greater than $+60$	10^{-10} and smaller	'No' reaction

Table 17.3a
The relationship between standard free energy and equilibrium constant for a reaction

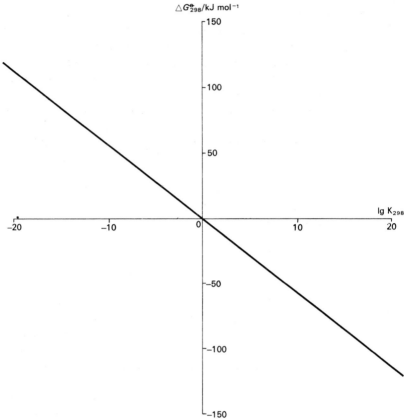

Figure 17.3a
Graph of $\Delta G_{298}^{\ominus}$ plotted against lg K_{298}.
(Since $\Delta G^{\ominus} = -2.303\ RT$ lg K, the gradient of this graph is $-2.303\ RT$. At 298 K, as here, its value is -5.71 kJ mol^{-1}.)

Questions

The questions in the *Students' Book* (simple numerical problems) at this point are intended to give students an opportunity to familiarize themselves with the relationship between $\Delta G^{\ominus}$ and K.

The answers are:
    1 $\Delta G_{298}^{\ominus} = -33.3$ kJ
    2 $\Delta G_{298}^{\ominus} = +61.1$ kJ
    3 $K_c = 10^{37}$

$\Delta G^{\ominus}$ and $\Delta H^{\ominus}$

Now that $\Delta G^{\ominus}$ has been defined and introduced there is an opportunity to clarify further one matter which was not explained completely to the students in Topic 7 (Energy changes and bonding). In Topic 7 we assumed that if a chemical reaction was strongly exothermic (ΔH negative) then the products of that reaction would be 'energetically stable', that is, K for the reaction would be large, and, correspondingly, if a chemical reaction was strongly endothermic (ΔH positive) then the products would be 'energetically unstable', that is, K would be small. It is necessary to hedge about this generalization (historically known as the Berthelot-Thomsen rule) with several provisos, of which the following are the most important:

1   The ΔH values (positive or negative) must be considerable in magnitude, for example more than 40 or 50 kJ mol^{-1}.

2   The temperature must not be too far removed from room temperature (298 K).

3   There are exceptions to this rule (some endothermic chemical processes are known which occur spontaneously at ordinary temperatures).

In the *Teachers' Guide* for Topic 7 it was mentioned that, as we now also see, it is $\Delta G^{\ominus}$ and not $\Delta H^{\ominus}$ that is related to K_c in the way we were assuming. Students can now investigate for themselves how far our assumptions were justified and also the necessity for making the three provisos (listed above), by noting how closely $\Delta H^{\ominus}$ and $\Delta G^{\ominus}$ values agree under various conditions.

By looking at $\Delta H^{\ominus}$ values for the various reactions listed in table 3 in the appendix to this topic in the *Students' Book*, students can be expected and encouraged to reach for themselves the following conclusions.

1   For very large or very small values of K ($\geqslant 10^{10}$ and $\leqslant 10^{-10}$ respectively) $\Delta H_{298}^{\ominus}$ and $\Delta G_{298}^{\ominus}$ are of roughly the same order of magnitude, so that for very strongly exothermic or endothermic compounds the value of $\Delta H_{f,298}^{\ominus}$ is a reasonable substitute value for $\Delta G_{f,298}^{\ominus}$.

2   When K has intermediate values the pattern is not at all clear and $\Delta H_{298}^{\ominus}$ and $\Delta G_{298}^{\ominus}$ can have widely differing values. Students may also spot that the differences are large when large volume changes, caused by the reaction or production of gases (and consequently large entropy changes), are involved in the reactions and that the differences are small (for gas reactions) when the volume changes (and entropy changes) are negligible or zero.

3   By studying the values of $\Delta H_T^{\ominus}$ and $\Delta G_T^{\ominus}$ at various temperatures (table 2 in the appendix to this topic in the *Students' Book*) students should also be able to conclude that any correspondence between $\Delta H_T^{\ominus}$ and $\Delta G_T^{\ominus}$ which there may be at room temperature has entirely disappeared at higher temperatures

(at least at room temperature for gas reactions $\Delta H^\ominus$ and $\Delta G^\ominus$ are of the same sign, even if their magnitudes differ widely). Students should not really be surprised at this since they have already noted:

    a   That $\Delta G^\ominus$ is dependent on temperature (this section).

    b   That at temperatures around room temperature and above, $\Delta H^\ominus$ is largely independent of temperature (section 17.2).

The general message is therefore: $\Delta G^\ominus$ and not $\Delta H^\ominus$ is a measure of the magnitude of K: $\Delta H^\ominus$ is related quite differently to K (as in section 17.2).

Entropy

At this point in the topic it is possible to investigate further the reasons for the differences between $\Delta H^\ominus$ and $\Delta G^\ominus$ in order to arrive at the important relationship

$$\Delta G^\ominus = \Delta H^\ominus - T\Delta S^\ominus$$

and the even more general one

$$\Delta G = \Delta H - T\Delta S$$

This could be done in several ways, of which the following outlines are examples.

    1   Starting from

$$\frac{\mathrm{d} \ln K_p}{\mathrm{d}\left(\dfrac{1}{T}\right)} = -\frac{\Delta H^\ominus}{R}$$

assume that $\Delta H^\ominus$ is independent of temperature and integrate:

$$R \ln K_p = \text{constant} - \frac{\Delta H^\ominus}{T} \quad \text{(as derived in 17.2)}$$

Compare this with the definition

$$\Delta G^\ominus = -RT \ln K_p \quad \text{(this section)}$$

so that

$$\Delta G^\ominus = \Delta H^\ominus - \text{constant} \times T$$

Name the 'constant' $\Delta S^\ominus$. Point out that the 'constancy' of $\Delta S^\ominus$ depends upon our assumption of the 'constancy' of $\Delta H^\ominus$

$$\therefore \Delta G^\ominus = \Delta H^\ominus - T\Delta S^\ominus$$

    2   For electrochemical cell reactions, for example, in the cell

$$Cu(s)|Cu^{2+}(aq) \vdots Ag^+(aq)|Ag(s),$$

investigate experimentally the way in which the value of $(\Delta H^{\ominus} - \Delta G^{\ominus})$ varies with T and hence conclude that

$$\Delta H^{\ominus} - \Delta G^{\ominus} = \text{constant} \times T$$

This could also be arrived at by using published data for gas reactions (see appendix table 2 for example) for $\Delta H_T^{\ominus}$ and $\Delta G_T^{\ominus}$. As before, the 'constant' is named $\Delta S^{\ominus}$, the standard entropy change for the reaction.

Teaching 'entropy' to advanced students
If the students meet $\Delta S^{\ominus}$ as a 'constant' in the equation

$$\Delta H^{\ominus} = \Delta G^{\ominus} - \text{constant} \times T,$$

and are told that $\Delta S^{\ominus}$ is called the standard entropy change for the reaction, they will, it is to be hoped, start asking questions about entropy, about which they certainly will have heard or read. Three questions they may well ask are:

1   How can I understand entropy? (What does it *mean*?)
2   How can I measure it?
3   How can I use it?

The problem presents itself therefore of deciding how far to go in trying to answer these three questions, since a considerable amount of work and a number of new ideas are involved in a discussion of entropy at this level. What follows is a necessarily brief account of some ways of trying to answer these questions, ending with a short discussion of the advantages and disadvantages of pursuing this in a school course.

1   What does entropy mean?
There are several ways of discussing this. One good way begins along the lines suggested by H. A. Bent,* and continues through the sequence (A) to (C).

A   Science, it has been said, begins in observation. The best way to become acquainted with birds, for example, is to look at some birds. A good way to become acquainted with entropy is to look at some entropies.

Substance	Molar entropy at room temperature and atmospheric pressure †
diamond	2.5
platinum	41.8
lead	64.8
laughing gas	220

---

* Bent, H. A. (1965) *The second law*, chapter 5, Oxford University Press, New York.
† The units of entropy will be considered later. They are the same as the units of heat capacity: joules per kelvin per mole.

These numbers suggest that entropy is related to hardness. Indeed, as a rule, hard, gem-like, abrasive, and refractory materials such as diamond, garnet, topaz, quartz, fused zirconia, silicon carbide, and boron nitride have small measured entropies; in these the individual atoms are bound to each other in nearly infinite, three-dimensional lattices by genuine chemical bonds that severely limit random thermal motions of atoms. On the other hand, soft substances, especially gases, usually contain large amounts of thermal disorder at room temperature and have, correspondingly, large measured entropies. Lamellar crystals are an interesting exception. While hard substances always have small entropies, not all soft substances have large entropies. Graphite, which is hard in two dimensions, but soft in the third, has an entropy at room temperature of only 5.61 entropy units per mole. Complex substances generally have larger entropies than simple substances of similar hardness. The entropies at room temperature of copper, sodium chloride, and zinc chloride, for example, are 33.5, 72.4, and 108 entropy units per gramme formula weight. Another example is sodium sulphate and its decahydrate; their entropies at 25 °C are, respectively, 149 and 593.

It is possible to become acquainted with entropies in an empirical way, as suggested by Bent. It is established that for every pure substance it is possible to determine (we shall discuss how later) a value for its standard molar entropy. The word molar indicates entropy per mole of the substance, and implies that entropy is an *extensive* property of substances (like heat capacity and volume) in that its magnitude depends on the amount of the substance considered.

B   If every pure substance has an entropy, then it must be possible to calculate values for the changes in the entropies of substances when a chemical reaction takes place. Various examples can be cited.

Example 1 : *Entropy changes on evaporation*. The standard molar entropies of liquid water and of water vapour are 69.9 and 188.7 respectively so that for the change

$$H_2O(l) \rightarrow H_2O(g)$$

$$\Delta S^{\ominus} = 188.7 - 69.9 = 118.8$$

Students could look at data such as that in table 17.3b, and conclude that for most liquids the molar entropy of evaporation is about the same (Trouton's rule). It is also possible by comparing $\Delta S^{\ominus}$, $\Delta H^{\ominus}$, and the boiling points of the liquids, to hint at how $\Delta S^{\ominus}$ was arrived at for the evaporation of these liquids. This point will be elaborated later.

Name	$\Delta S_{vap}/J\ K^{-1}\ mol^{-1}$	Boiling point/K	$\Delta H_{vap}/J\ mol^{-1}$
Hydrocarbons			
methane	73.6	112	8180
ethane	80.2	184	14720
propane	82.8	231	19250
butane	82.0	273	22380
pentane	83.3	309	25770
hexane	84.5	342	28870
heptane	85.4	371	31690
octane	86.2	399	34350
nonane	87.5	424	36900
benzene	87.0	353	30750
toluene	83.7	384	31970
naphthalene	70.3	591	40540
Halogenated compounds			
chloroethane	86.6	285	24690
trichloromethane	87.9	334	29370
tetrachloromethane	85.8	349	30000
iodobenzene	85.4	461	39410
Hydroxy-compounds			
water	109	373	40580
methanol	105	338	35270
ethanol	110	351	38580
propanol	110	370	40580
butan-1-ol	113	390	43540
pentan-1-ol	127	411	51900
hexan-1-ol	132	430	56480
heptan-1-ol	114	449	51040
Carboxylic acids			
formic acid	61.5	373	22930
acetic acid	62.3	391	24310
propionic acid	69.0	414	28330
butanoic acid	96.2	436	41860
pentanoic acid	95.4	460	43940
Miscellaneous compounds			
carbon disulphide	84.1	319	26780
diethyl ether	87.0	307	26690
acetone	94.6	329	31090

Table 17.3b
Entropy changes on vaporization at constant pressure

Example 2. Table 17.3c shows values of the entropy changes for the various types of reaction listed in table 3 of the appendix to this topic in the *Students' Book*.

Reaction	$\Delta S^{\ominus}_{298}$/J K^{-1} mol^{-1}	$T\Delta S^{\ominus}_{298}$/kJ mol^{-1}
Type of reaction: gas + gas → gas		
$CH_4(g) + 2O_2(g) \rightarrow CO_2(g) + 2H_2O(g)$	−5.61	−1.67
$H_2(g) + Cl_2(g) \rightarrow 2HCl(g)$	+19.7	+5.86
$CCl_4(g) + 2H_2O(g) \rightarrow CO_2(g) + 4HCl(g)$	+275	+82.0
$N_2O_4(g) \rightarrow 2NO_2(g)$	+177	+52.7
$N_2(g) + 3H_2(g) \rightarrow 2NH_3(g)$	−198	−59.0
$C_2H_4(g) + H_2(g) \rightarrow C_2H_6(g)$	−122	−36.4
Type of reaction: solid + gas → solid		
$CaO(s) + CO_2(g) \rightarrow CaCO_3(s)$	−160	−47.7
$2Cu(s) + O_2(g) \rightarrow 2CuO(s)$	−188	−56.1
Type of reaction: solid + gas → gas		
$C(s) + O_2(g) \rightarrow CO_2(g)$	+2.80	+0.8
$C(s) + H_2O(g) \rightarrow CO(g) + H_2(g)$	+135	+40.2
Type of reaction: solid + gas → solid + gas		
$MgO(s) + CO(g) \rightarrow Mg(s) + CO_2(g)$	+21.7	+6.7
$ZnO(s) + CO(g) \rightarrow Zn(s) + CO_2(g)$	+13.4	+4.2
Type of reaction: solid + solid → solid		
$Cu(s) + S(s) \rightarrow CuS(s)$	+1.38	+0.4
$2Al(s) + Cr_2O_3(s) \rightarrow Al_2O_3(s) + 2Cr(s)$	−39.3	−11.7
$FeO(s) + Fe_2O_3(s) \rightarrow Fe_3O_4(s)$	−11.1	−3.3

Table 17.3c
Entropy changes for the reactions listed in *Students' Book* appendix table 3

From a study of the entropy changes accompanying these reactions it is possible for students to reach the following conclusions.

a   In certain types of chemical reaction, notably reactions in which there is a change in the number of moles of gases involved, the numerical values of $\Delta S^{\ominus}_{298}$ are often quite large. These are the very same reactions as those for which $\Delta H^{\ominus}_{298}$ and $\Delta G^{\ominus}_{298}$ differ markedly. If there is an *increase* in gaseous volume on reaction, $\Delta S^{\ominus}_{298}$ is *positive*, and if a *decrease* in volume, $\Delta S^{\ominus}_{298}$ is *negative*. In both of these cases $T\Delta S^{\ominus}_{298}$ is also numerically quite large, and there is therefore a sizeable difference between $\Delta H^{\ominus}_{298}$ and $\Delta G^{\ominus}_{298}$.

b   In chemical reactions for which $\Delta H^{\ominus}_{298}$ and $\Delta G^{\ominus}_{298}$ do not differ very greatly, there is usually little difference between the entropies of the reactants and those of the products, i.e. $\Delta S^{\ominus}_{298}$ is small, and so also therefore is $T\Delta S^{\ominus}_{298}$.

Students can in this way get some general idea empirically of what sort of entropy changes they will expect to find accompanying chemical reactions of various kinds.

Example 3: Entropy values and temperature.

Standard molar entropy, $S^{\ominus}/J\ K^{-1}\ mol^{-1}$

Gas	298 K	500 K	1000 K	1500 K	2000 K
H_2	130	146	166	179	188
N_2	192	207	228	242	252
O_2	205	221	244	258	269
F_2	199	220	245	260	270
Cl_2	223	241	267	282	293
Br_2	245	264	290	305	317
I_2	261	280	306	321	332
CH_4	186	207	248	280	306
NH_3	193	212	246	271	290
H_2O	189	206	233	251	265
HF	174	189	209	222	231
HCl	187	202	223	236	246
HBr	199	214	235	248	259
HI	207	222	243	257	268
CO	198	213	235	249	259
CO_2	214	235	269	292	309
SO_2	248	270	305	328	344

Table 17.3d
Standard entropies of gases at various temperatures. All values are at 1 atm pressure and are corrected for deviations from the perfect gas laws. *After Caldin, E. F.* (1958) An introduction to chemical thermodynamics. *Oxford University Press.*

Examination of a table such as table 17.3d soon leads to the conclusion that the entropies of substances increase as the temperature increases. It is also noticeable that they all increase more or less 'in step', that is by similar amounts for the same temperature rise.

C   The entropy values encountered in (A) can be understood as a measure (in fact a scaled down logarithmic measure) of the number of ways in which the energy of the molecules, ions, and other particles, can be shared out or 'spread' over the available energy levels. This can lead to a very useful qualitative means of explaining how it comes about that gases have higher entropies than liquids, which in turn have higher entropies than solids; or why for a given molecular weight more complex molecules have higher entropies than simpler molecules (see table 17.3e below).

Molecular weight ≈ 20		Molecular weight ≈ 40		Molecular weight ≈ 80		Molecular weight ≈ 120	
Ne	146	Ar	155	Kr	164	Xe	170
HF	174	F_2	203	Cl_2	223	BrCl	240
H_2O	189	CO_2	214	CS_2	238	NOBr	273
NH_3	193	H_2O_2	233	SO_3	256	$COCl_2$	281
CH_4	186			CH_2Cl_2	271	$CHCl_3$	296
				N_2O_4	304		

Table 17.3e
Entropies of gases in order of increasing molecular complexity. (Values of $S_{298}^{\ominus}$ in J K^{-1} mol^{-1}.)

To do this effectively it is necessary for students to understand clearly that

a   the greater the number of ways in which it is possible to distribute a given amount of energy among a collection of particles such as molecules and ions, the higher will be their entropy;

b   the number of ways in which energy can be distributed will depend on (1) the types and (2) the numbers of *energy levels* over which the energy can be distributed.

1   The energies (rotational, vibrational, translational, electronic) of molecules are *all* quantized (though in the case of translational motion the levels are so close together that we can usually regard them as a continuum). Gases have translational energy which may be distributed over a range of energy levels, and in this they differ from solids in which the particles have vibrational and rotational energies but no translational energy. The entropies of gases are therefore correspondingly greater than those of solids. Liquids form an intermediate case but are more like solids than like gases. It is possible to calculate in simple cases the various contributions to the total entropies of gases made by spreading the energy over the translational, rotational, and vibrational energy levels, as the following examples show.

	Translational entropy (1 atm)	Rotational entropy	Vibrational entropy	Total entropy (1 atm)
$N_2O(g)$	156	54	10	220
$H_2O(g)$	145	44	0	189

Table 17.3f
The various contributions to the total entropy of gases. (Values are in J K^{-1} mol^{-1} at 298 K.) The small contributions that vibrational degrees of freedom make to the entropies of water vapour and dinitrogen monoxide at room temperature arise because of the fact that at this temperature the vibrational degrees of freedom of these molecules are not easily excited. *After Bent, H. A.* (1965) *The second law, Oxford University Press, New York.*

2 The *numbers* of energy levels over which the energy of a substance may be distributed depend on the temperature of the substance. At higher temperatures, more molecules are moving faster and higher translational energy levels can become more densely populated (see figure 17.3b), so that there is an increase in the number of ways in which the energy can be shared among the accessible energy levels and consequently an increase in entropy.

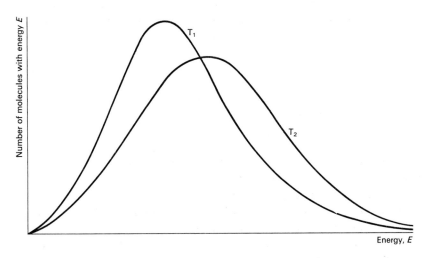

Figure 17.3b
The distribution of energy amongst molecules at a temperature T_1, and at a higher temperature T_2.

The entropies of substances therefore increase the hotter they become (see figure 17.3c for example or table 17.3d).

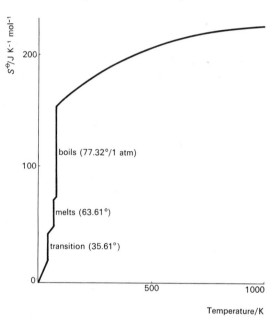

Figure 17.3c
Standard entropy of nitrogen, at one atmosphere pressure, as a function of temperature.
After Caldin, E. F. (1958) An introduction to chemical thermodynamics, *Oxford University Press*

The entropy of a substance is defined as $S = k \ln W$ where W is the number of ways in which the energy of a substance can be distributed over all the accessible energy levels and k is a scaling factor known as Boltzmann's constant. The value of W is frequently enormous, so great in fact that it is impossible to count it. For example, W for half a mole (6 g) of diamond, which has an entropy of 1.25 J K^{-1} has a value $10^{0.4 \times 10^{23}}$. The universe itself is only about 3×10^{23} microseconds old, so that counting a number this large is out of the question.

The *direction of change* in natural processes can also be understood simply and qualitatively in terms of relative numbers of ways of distributing energy. If before reaction a system has entropy S_1 and there are W_1 ways of distributing the energy within the system, and if after reaction the system has entropy S_2

and the number of ways is now W_2, then

$$\Delta S = S_2 - S_1 = k \ln W_2 - k \ln W_1$$

$$\therefore \Delta S = k \ln \frac{W_2}{W_1}$$

It follows that ΔS is positive only if $W_2 > W_1$ (since k is positive). That is, the entropy increases only if, as a result of reaction, the number of energy levels accessible in state 2 is greater than the number accessible in state 1. Since reactions naturally occur in such a way that the most probable distribution of energy is achieved (greatest number of ways of distributing the energy), natural processes are accompanied by an increase in entropy. This is, however, a very general statement and needs further qualification as in part (3), How can entropy be used? It does, however, help to explain why it is that reactions in which large quantities of gases are released such as

$$6SOCl_2(l) + FeCl_3, 6H_2O(s) \rightarrow FeCl_3(s) + 6SO_2(g) + 12HCl(g)$$

take place naturally at room temperature, even though they are strongly endothermic; they are sometimes said to be 'entropy driven'.*

Disorder. It is often suggested that entropy can be pictured as the disorder or randomness or chaos in a system. This is an attractive idea, and it has the virtue of intuitive simplicity, but there are several pitfalls to watch out for in using it. One is that 'disorder' or 'randomness' or 'chaos' immediately suggests to us *spatial* disorder. In a number of cases this is useful, but, always, the more general and fundamental idea is that of the number of ways in which energy can be shared out among the accessible energy levels. There is a temptation to load up the term 'disorder' with this special meaning which is *not* intuitively obvious. Another difficulty with disorder is that there are cases in which it is difficult to decide how one could ever tell, except by an independent knowledge of the entropies concerned, which of two systems is more disordered. For gases of a given molecular weight, for example (see table 17.3e), is there more disorder in a gas with diatomic molecules or in one with tetra-atomic molecules and if so, how do you tell?

There is no doubt, however, that, except for the purists among us, the disorder idea can give students a very useful introductory insight into the meaning and nature of entropy and entropy changes. For those students for whom probability distributions, quantization of energy levels, and the like, are too much to bear, the disorder approach, even if it has its disadvantages, may very well be a useful way of thinking about entropy.

*Matthews, G. W. J. (1966) 'A demonstration of spontaneous endothermic reaction', *J. Chem. Educ.* **43**, 476.

2   How can entropy be measured?

Here it is really necessary to present the definition of an entropy change as

$$\Delta S = \frac{q_{\text{reversible}}}{T}$$

or, in differential notation,

$$dS = \frac{\delta q_{\text{reversible}}}{T}$$

It may be possible for students to understand this by considering a special case, namely the isothermal evaporation of a liquid. Students may have noticed, when studying table 17.3b, with a little help, that in this case

$$\Delta S_{\text{evaporation}} = \frac{\Delta H_{\text{evaporation}}}{T_{\text{b}}},$$

where T_{b} is the boiling point of the liquid. The evaporation of a liquid at constant temperature and pressure is a useful simple example of a *thermodynamically reversible* process. If the liquid is allowed to evaporate isothermally it draws in from its surroundings the molar heat of evaporation, ΔH_{vap}. (The entropy gain of the substance, ΔS_{system}, equals $\dfrac{\Delta H_{\text{vap}}}{T}$ and the entropy loss of the surroundings equals $\dfrac{\Delta H_{\text{vap}}}{T}$ so that $\Delta S_{\text{total}} = 0$, the condition for a reversible process.) If the liquid is now condensed isothermally, the molar heat of evaporation is returned from the substance to its surroundings so that the situation is exactly restored to what it was at the beginning (another mark of thermodynamic reversibility).

If substances are warmed up under thermodynamically reversible conditions (not through phase changes but the ordinary raising of the temperature of a substance in one phase) the entropy increase of the substance on warming up is

$$\frac{\delta q_{\text{reversible}}}{T}$$

for each small increase in thermal energy, $\delta q_{\text{reversible}}$, which the substance receives. If the warming up takes place at constant pressure, say one atmosphere, then we can write the increase in the energy of the substance dH, so that the entropy increase of the substance,

$$dS = \frac{dH}{T}$$

Putting it another way

$$\frac{dS}{dH} = \frac{1}{T}$$

Using the standard procedure in simple calculus called 'function of a function' we can say

$$\frac{dS}{dT} \cdot \frac{dT}{dH} = \frac{1}{T}$$

or

$$\frac{dS}{dT} = \frac{1}{T} \frac{dH}{dT} \quad \text{(equation 1)}$$

Remembering that the molar heat capacity at constant pressure C_p is the rate at which one mole of a substance can gain enthalpy per kelvin (that is, the derivative of the enthalpy of the substance with respect to temperature, $\frac{dH}{dT}$), equation 1 becomes

$$\frac{dS}{dT} = \frac{1}{T} \cdot C_p \quad \text{(equation 2)}$$

Calling S_0 and S_T values of the entropy at 0 and T K respectively, we can write this differential equation (2) in an integral form:

$$S_T - S_0 = \int_0^T \frac{C_p}{T} \, dT$$

For one mole of a substance (if gaseous at a fixed pressure of 1 atm) these entropies are the standard values $S_T^{\ominus}$ and $S_0^{\ominus}$. To evaluate $S_T^{\ominus} - S_0^{\ominus}$, then, all we have to do is plot values of $\frac{C_p}{T}$ for various values of T and estimate the area under the graph of $\frac{C_p}{T}$ against T between the limits of 0 K and T K.

A typical example is the determination of the entropy of silver chloride.* Figure 17.3d shows a plot of C_p for silver chloride against T.

---

*Eastman, E. D. and Milner, R. T. (1933) 'Entropy of a crystalline solution of silver bromide and silver chloride in relation to the Third Law of Thermodynamics', *J. Chem. Phys.* **1**, 444.

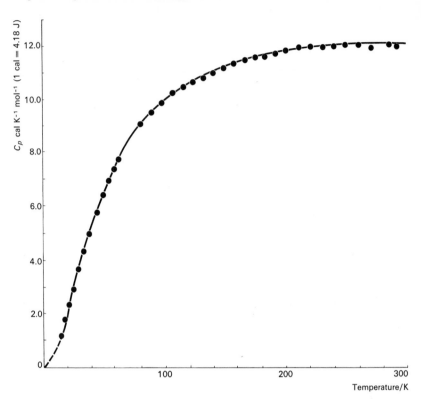

Figure 17.3d
Graph of C_p against T for silver chloride.
After Caldin, E. F. (1958)

Values of C_p were determined for silver chloride by calorimetry at temperatures down to about 15 K. Below 15 K the graph has been extrapolated (the dotted part) according to the Debye theory of solids which suggests that, in this temperature region, $C_p \propto T^3$.

The value of $\int_0^T \dfrac{C_p}{T}\,dT$ is found as the area under a plot of $\dfrac{C_p}{T}$ against T as in figure 17.3e.

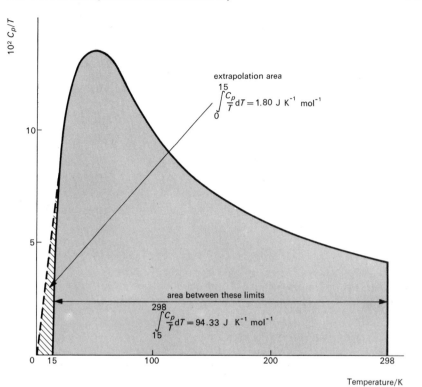

Figure 17.3e
Graph of C_p/T against T for silver chloride.
After Caldin, E. F. (1958)

Since C_p decreases with T, and more rapidly than T, $\dfrac{C_p}{T}$ tends to zero as T tends to zero, and not to ∞ as it would if C_p were finite at 0 K.

Standard entropies of solids, relative to their values at 0 K can be evaluated in this way to within about 0.8 J K^{-1} mol^{-1}. The extrapolation to zero introduces no great uncertainty. The entropies of pure liquids and gases can also be found by similar means by including the appropriate $\dfrac{\Delta H}{T}$ terms to allow for melting and evaporation. Figure 17.3c shows how the standard entropy of nitrogen increases with temperature, including increases due to phase changes.

Since, therefore, $S_T^{\ominus} - S_0^{\ominus}$ can be determined, it remains to ask whether we may assume, as Nernst did, that $S_0^{\ominus}$ is the *same* for all substances or, as Planck did, that $S_0^{\ominus}$ is *zero* for all pure, perfectly crystalline substances. These two assumptions are sometimes called the Third Law of Thermodynamics (weak and strong forms respectively). If the Third Law is assumed, then $S_0^{\ominus}$ may be arbitrarily or actually (depending upon which form of the Third Law is assumed) equated to zero, and the value of $\int_0^T \dfrac{C_p}{T}\,dT$ taken to be $S_T^{\ominus}$ for the substance. This is why $S_T^{\ominus}$ is sometimes called the 'Third Law entropy'. The standard entropies of elements are given in the *Book of Data*.

Calculation of entropies from statistical mechanics. The standard entropy for some pure substances relative to their values at 0 K can also be calculated from statistical mechanics and spectroscopic data. The treatment is beyond the scope of this summary except to say that the calculations depend on the definition of entropy in statistical mechanics as

$$S = k \ln W$$

first proposed by Planck in 1906 and later by Boltzmann. There is a readable account of this method in the first few chapters of W. S. Rushbrooke's *Introduction to statistical mechanics* (Oxford University Press, 1949).

3   How can entropy be used?

Students will probably have come across the dictum first formulated by Clausius (1865) that the entropy of the universe tends to increase.* This statement is in many ways the most general and simple form of the Second Law of Thermodynamics. What does it mean and how can it help chemists to understand the direction of chemical change?

Having spent some time discussing what is meant by entropy, we now have to consider briefly what is meant by 'universe' in this statement. It was originally intended to mean the whole of the physical universe, and, from a strictly rigorous point of view, this is so. It is possible, however, to try to set up little 'sub-universes' to which the second law in this form applies within the total universe, by *isolating* portions of the whole, so that whatever happens in the isolated portion does not in any way affect the rest of the universe. One theoretical problem this raises is that even if one does actually succeed in isolating completely a portion of the universe and one is outside it, one cannot possibly know what is happening inside it, since to find out one has to withdraw information in the form of light or electrical signals from the portion and, if it is truly isolated, one cannot do so. In practical terms, however, the departure from true isolation introduced by observing instruments such as thermometers placed in the 'isolated' portion is negligible. In this way the idea of the 'universe' of an event such as a chemical reaction can be introduced.

* 'Die Entropie der Welt strebt einem Maximum zu'.

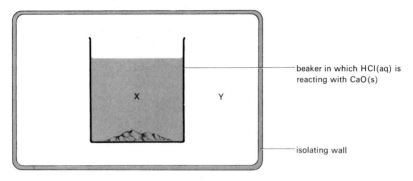

Figure 17.3f

Figure 17.3f shows the universe of the reaction between hydrochloric acid and calcium oxide which is going on in the *system* X. The *surroundings* Y in practical terms we can usually regard as 'the laboratory' or perhaps 'the fume cupboard'.

Clausius' statement can be applied to this 'universe' as follows: $\Delta S_{total} > 0$, that is, the changes which are taking place within this universe are accompanied by an overall increase in entropy.* Since

$$\Delta S_{total} = \Delta S_{system} + \Delta S_{surroundings},$$

it follows that $\Delta S_{system} + \Delta S_{surroundings} > 0$.

Suppose that the chemical process is exothermic and that the heat evolved, $-\Delta H$, is transferred to the surroundings. Then if the surroundings (the laboratory) are large compared with the system (the beaker) so that the temperature of the surroundings does not change appreciably, the entropy gained by the surroundings is $\dfrac{-\Delta H}{T}$. Therefore, $\Delta S_{system} - \dfrac{\Delta H}{T} > 0$ for the ongoing process.

Usually, at room temperatures, the entropy increase of the surroundings, $-\dfrac{\Delta H}{T}$, is much greater than the entropy increase (or decrease) of the system, so that, overall, the entropy of system and surroundings taken together increases. This is simply an expression of the well-known fact that most of the reactions which occur spontaneously are exothermic. In cases of spontaneous endothermic reactions (such as the reaction between thionyl chloride and iron(III) chloride hexahydrate) ΔS_{system} is here so large that, even though $\Delta S_{surroundings}$ is negative (ΔH being positive), ΔS_{total} is still positive.

---

*There is a limiting case, that of a thermodynamically reversible process, for which $\Delta S_{total} = 0$ but ΔS_{total} is never less than zero.

Why, at ordinary temperatures, should the entropy change of the surroundings usually be the dominant term in the above equation? The reason lies in the fact that the surroundings are usually much larger than the system and so the heat energy transferred from system to surroundings can be distributed in very many more ways in the surroundings than it was distributed in the system.

It is possible to develop a general qualitative explanation of the direction of spontaneous chemical change in terms of the entropy changes involved, along the following lines.

Spontaneous exothermic processes

Suppose for a spontaneous exothermic process

$$\Delta H = -a \, \text{J mol}^{-1} \, (a, b, c \dots \text{as used here are all assumed to be positive numbers})$$

Then $\Delta S_{\text{surroundings}} = +\dfrac{a}{T} \, \text{J K}^{-1} \, \text{mol}^{-1}$

Since for a spontaneous process, $\Delta S_{\text{total}} > 0$ it follows that

$$\frac{a}{T} + \Delta S_{\text{system}} > 0$$

$$\therefore \Delta S_{\text{system}} > -\frac{a}{T}$$

This implies that the entropy of the system *can* decrease, that is, ΔS_{system} can be negative, but it cannot be less than $-\dfrac{a}{T}$. If ΔS_{system} were less than $-\dfrac{a}{T}$ there would not be a net increase of entropy overall. $-\dfrac{a}{T}$ therefore sets a lower limit for the value of ΔS_{system} for a spontaneous process. Taking a numerical example, if $\Delta H = -30$ kJ at 300 K then ΔS_{system} cannot be less than but may be down to $-\dfrac{30\,000}{300}$ or -100 J K^{-1} for a spontaneous process. If $\Delta H = -0.3$ kJ, ΔS_{system} cannot be less than $-\dfrac{300}{300}$ or -1 J K^{-1}. This then makes clearer why, if ΔH is large and negative, reactions, even those reactions for which ΔS_{system} is negative, are likely to take place.

Spontaneous endothermic processes
Suppose a process for which $\Delta H = +b$ J mol^{-1}.

Then $\Delta S_{\text{surroundings}} = -\dfrac{b}{T}$ J K^{-1} mol^{-1}

and $-\dfrac{b}{T} + \Delta S_{\text{system}} > 0$

or $\Delta S_{\text{system}} > \dfrac{b}{T}$

For a spontaneous reaction, then, the entropy change in the system must be greater than $+\dfrac{b}{T}$ J K^{-1}. Again taking numerical examples, if

$$\Delta H = +30 \text{ kJ}, \Delta S_{\text{system}} > \frac{30\ 000}{300} \text{ or } 100 \text{ J K}^{-1}$$

$$\Delta S_{\text{surroundings}} = -100 \text{ J K}^{-1}$$

If $\Delta H = +0.3\text{kJ}, \Delta S_{\text{system}} > \dfrac{300}{300}$ or 1 J K^{-1}

$$\Delta S_{\text{surroundings}} = -1 \text{ J K}^{-1}$$

This means that endothermic processes will take place spontaneously only if the system is one in which a sufficiently large entropy increase can take place to outweigh the entropy *decrease* of the surroundings which is accompanying such a process.

Variation of temperature
So far we have considered cases at one temperature (say 300 K). We now consider how $\Delta S_{\text{surroundings}}$ and ΔS_{system} vary with T.

Spontaneous exothermic processes – Taking the change for which $\Delta H = -a$ J

$\Delta S_{\text{system}} > -\dfrac{a}{T}$

The variation of $\Delta S_{\text{surroundings}}$ and the *lower limit of* ΔS_{system}, that is, the lowest value ΔS_{system} can have if ΔS_{total} is to be positive at that temperature, is shown in figure 17.3g.

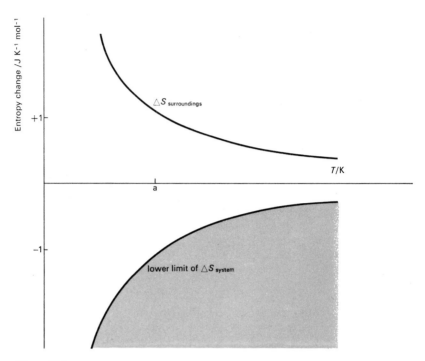

Figure 17.3g
Graph of entropy change against temperature for an exothermic change.

The assumption is made that ΔH is approximately independent of temperature over a fairly wide range. This is usually a reasonable assumption to make.

Figure 17.3g shows the lower limit of the value of ΔS_{system} at various temperatures. At any temperature, for spontaneous reaction $\Delta S_{total} > 0$ and for this to be so, ΔS_{system} must have some value above the lower curve. At low temperatures ($\approx a$ K) ΔS_{system} can be quite large and negative. At high temperatures $\Delta S_{surroundings} \rightarrow 0$, so that ΔS_{system} cannot have (in the limit) a negative value.

Spontaneous endothermic processes – Now consider the change for which $\Delta H = +b$ J (figure 17.3h).

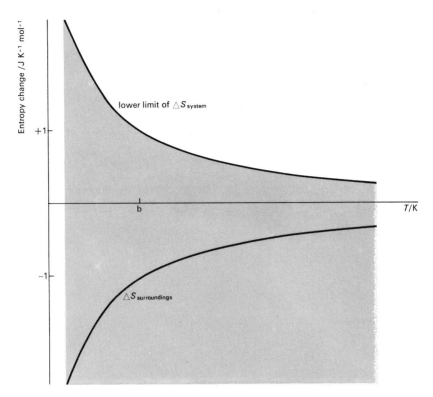

Figure 17.3h
Graph of entropy change against temperature for an endothermic change.

At any temperature $\Delta S_{\text{total}} > 0$ and, since $\Delta S_{\text{surroundings}}$ is given by the lower curve, ΔS_{system} must lie *above* the upper curve (for a spontaneous process). At low temperatures this requires ΔS_{system} to be quite large and positive. At high temperatures, ΔS_{system}, while still positive, can have much lower values and in the limit can approach zero.

This approach applied to some examples
1   The reaction $\frac{1}{2}H_2(g) + \frac{1}{2}Cl_2(g) \rightarrow HCl(g)$.

As can be seen in figure 17.3j, this reaction proceeds to completion at all temperatures up to 2000 K and by extrapolating by eye would apparently not be reversed until the temperature reached about 20 000 K.

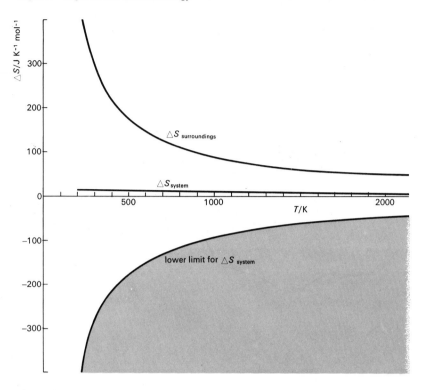

Figure 17.3j
Graph of entropy change plotted against temperature for the system
$\frac{1}{2}H_2(g) + \frac{1}{2}Cl_2(g) \rightarrow HCl(g)$
and its surroundings. $\Delta H^{\ominus}_{298} = -92.5$ kJ.

2   The reaction $C(s) + H_2O(g) \rightarrow CO(g) + H_2(g)$.

Figure 17.3k shows the variation of $\Delta S^{\ominus}_T$ for this reaction. At temperatures below about 900 K this reaction does not take place but above this temperature the value of $\Delta S^{\ominus}_{\text{system}}$ exceeds the lower limit of ΔS_{system} (the curve obtained by plotting $+\dfrac{\Delta H^{\ominus}_{298}}{T}$ against T) so that above about 900 K this reaction becomes 'feasible'.

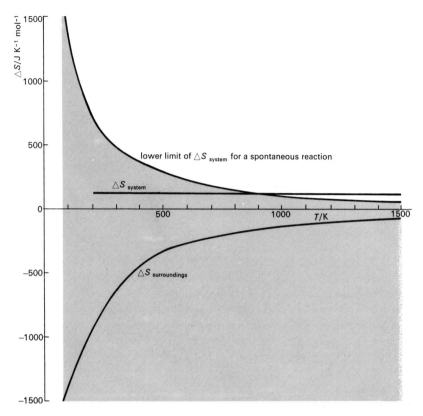

Figure 17.3k
Graph of entropy change plotted against temperature for the system
$$C(s) + H_2O(g) \rightarrow CO(g) + H_2(g)$$
and its surroundings. $\Delta H^{\ominus}_{298} = +131$ kJ.

In using entropies to forecast the likely direction of chemical change we have used the inequality

$$\Delta S - \frac{\Delta H}{T} > 0.$$

Multiplying through by $-T$, we get

$$\Delta H - T\Delta S < 0.$$

Chemists prefer a single symbol for this composite expression on the lefthand side: they write instead ΔG.

This, then, is the sort of work involved in discussing with advanced school students the understanding, measurement, and use of entropy in chemistry. Whether or not the effort involved is worth the return in terms of students' increased understanding of the factors affecting the direction of chemical change will be a matter of personal judgment for individual teachers, bearing in mind the abilities of their students, the time available, whether or not entropy will be covered in an advanced school physics course, and so on. It is certainly desirable on general educational grounds that sixth form science students (perhaps *all* sixth form students?) should have some understanding of entropy. From the point of view solely of understanding the direction of chemical change, however, they can get a long way in terms of equilibrium constants and $\Delta G^{\ominus}$. The choice of what to do is therefore left open to the teacher in this course. Accordingly the work on entropy appears in the *Teachers' Guide* only and does not appear in the *Students' Book*.

Using $E^{\ominus}$ values to forecast the direction of chemical change

The main objective of this section is to show students how to calculate equilibrium constants indirectly. We now take up, therefore, some ideas first introduced to students in Topic 16 (Some d-block elements), about the use of standard redox potentials ($E^{\ominus}$ values) for predicting the likely direction of chemical change. It is suggested that students may need to revise section 16.1 before continuing with this sub-section.

The starting point takes the form of a set of revision questions whose answers require a knowledge of how to use $E^{\ominus}$ values predictively. The questions are in the *Students' Book*. The predictions can be tested by means of some simple test-tube experiments (experiment 17.3a).

Experiment 17.3a
To test predictions about some redox reactions

Each student or pair of students will need:
    Rack of test-tubes
    Spatula
access to:
    Bromine water
    Iron(II) ammonium sulphate solution
    Potassium hexacyanoferrate(III) solution
    Potassium hexacyanoferrate(II) solution
    (or potassium thiocyanate solution)
    Potassium iodide solution
    Potassium chloride solution
    Tin(II) chloride
    Iron(III) alum, iron(III) sulphate, or iron(III) chloride
    Potassium permanganate solution

Potassium bromide solution
Tetrachloromethane
Zinc dust
Copper powder
Dilute sulphuric acid solution
Approximately 2M ammonia solution

Full details of the procedure for this experiment are given in the *Students' Book*.

Discussion of results
The *Students' Book* asks questions which are intended to elicit from students:
   1   Whether their predictions using $E^{\ominus}$ values were borne out in practice by experimental results.
   2   An explicit formulation of the rule they were using in this predictive use of $E^{\ominus}$ values. The *Students' Book* takes the example of the reduction by zinc dust as an illustration, and table 17.3g summarizes the rule for using $E^{\ominus}$ values predictively.

$E^{\ominus}$	K_c for cell reaction
$+0.4$ V or larger	very large
-0.4 V or less	very small

Table 17.3g
Cell e.m.f.s and K_c for cell reactions

A glance at the values of $E^{\ominus}_{298}$ and $K_{c,298}$ in table 4 in the appendix to this topic in the *Students' Book* shows that this rule is a reasonable working guide to the direction of chemical reactions.

More precise relationship between $E^{\ominus}_{cell}$ and K_c
So far we have made only a qualitative use of standard redox potentials. The next suggested step is to use the Nernst equation for a metal/ion half-cell to deduce the more general expression for a cell and then note to what this expression reduces when the cell is short circuited and 'runs down' to equilibrium. There is a fairly full treatment of this in the *Students' Book*, ending with a quantitative relationship between $E^{\ominus}_{cell}$ and K_c, namely

$$E^{\ominus}_{cell} = \frac{2.3\,RT}{zF}\lg K_c.$$

In this expression the units of R are $J\ K^{-1}\ mol^{-1}$; R is therefore $8.31\ J\ K^{-1}\ mol^{-1}$. With $F = 96\ 500$ C, this gives $E^{\ominus}$ in volts.

Students' Book appendix table 4 lists some equilibrium constants for a selection of cell reactions. These K_c values have been calculated from the corresponding $E^{\ominus}_{\text{cell}}$ values using the Nernst expression.

Questions in the *Students' Book*
The questions which are asked at this point in the *Students' Book* are intended to enable students to familiarize themselves with the use of this equation. The answers are

1   Cell equilibrium reaction:
$$Cu(s) + Br_2(aq) \rightleftharpoons Cu^{2+}(aq) + 2Br^-(aq)$$
2   $K = 2 \times 10^{25} \text{ mol}^2 \text{ dm}^{-6}$

Experiments 17.3b and 17.3c
Students may naturally wish to test any indirect method of determining equilibrium constants that is suggested to them against the results which can be obtained by a direct method. The following pair of experiments affords an opportunity for them to do this. In experiment 17.3b students find what the equilibrium concentrations of an equilibrated system would be by a graphical treatment of cell e.m.f. measurements. In experiment 17.3c equilibrium concentrations are directly determined by titration. The values of K obtained by each of these methods can then be compared and both can also be compared with the value of K calculated from standard redox potentials.

Experiment 17.3b
Finding the equilibrium constant for the reaction
$Ag^+(aq) + Fe^{2+}(aq) \rightleftharpoons Fe^{3+}(aq) + Ag(s)$ by e.m.f. measurements
on the cell $Pt\,|\,Fe^{2+}(aq),\ Fe^{3+}(aq)\,|\,Ag^+(aq)\,|\,Ag$

Each student or pair of students will need
0.10M iron(III) nitrate in 0.05M nitric acid
   (approximately 5 cm³ of concentrated HNO_3 per cubic decimetre of solution)
0.20M iron(II) sulphate
0.20M barium nitrate
0.40M silver nitrate
Ammonium nitrate/gelatine salt bridge
Platinum electrode (foil or wire)
Silver electrode (foil or wire)
Valve voltmeter, pH meter, or potentiometer. The sensitivity of the meter should be such that e.m.f. can be determined to within one or two millivolts.

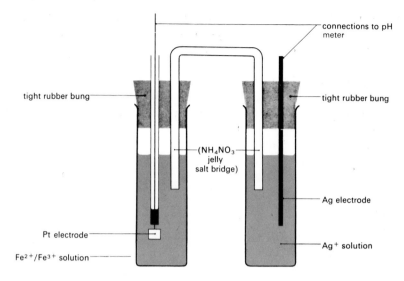

Figure 17.3l

Preparation of the $Fe^{2+}(aq)/Fe^{3+}(aq)$ half-cell

Equal quantities of 0.20M iron(II) sulphate and 0.20M barium nitrate are mixed and the precipitate allowed to settle for a few minutes in a stoppered vessel. The supernatant solution* is 0.10M in iron(II) nitrate. Equal volumes of the 0.1M iron(II) nitrate and iron(III) nitrate solutions are now mixed and transferred into the $Fe^{2+}(aq)/Fe^{3+}(aq)$ half-cell. The latter is properly sealed with a rubber bung fitted with the platinum electrode and the salt bridge, to avoid any air oxidation of the iron(II).

Preparation of the salt bridges

A ten per cent solution of gelatin or agar in concentrated (but not quite saturated) ammonium nitrate solution is brought to the boil and then immediately injected conveniently by means of a plastic syringe into the specially prepared glass U-tubes. The U-tubes are left to stand, preferably overnight, until the jelly has set.

Alternatively, thin strips of filter paper moistened with saturated ammonium nitrate solution may be jammed between the rubber bungs and glass walls of the half-cells and used in place of the jelly salt bridges.

---

*Small quantities of barium sulphate may remain suspended in the solution and will not affect the results adversely.

Procedure

The silver half-cell is set up as shown in the diagram, 0.4M $AgNO_3$ solution being used. The e.m.f. and polarity of the electrodes are recorded. Further measurements are carried out with silver nitrate solutions of molarities 0.20, 0.10, 0.050, 0.025, prepared by successive dilution of the initial 0.40M solution. If a pH meter is used for the measurements, it is necessary to change over the connectors to the cell at some stage during the measuring series.

Treatment of results

Students should note down their results as they obtain them in a table similar to that suggested in the *Students' Book*. The e.m.f. values measured should be plotted against $-\lg [Ag^+(aq)]$. A plot similar to that in figure 17.3m is obtained.

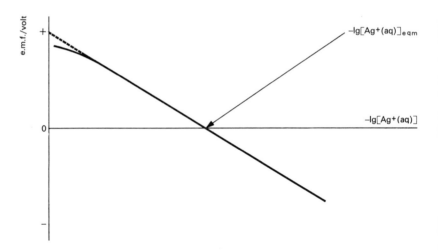

Figure 17.3m
A typical plot of the cell e.m.f. against values of $-\lg[Ag^+(aq)]$.

At low concentrations of silver ion the relationship between e.m.f. and $-\lg [Ag^+(aq)]$ is linear; at higher concentrations ($[Ag^+(aq)] > 0.1M$) some departure from linearity is usually observed.

The point at which the curve crosses the $-\lg [Ag^+(aq)]$ axis (that is, when the cell e.m.f. is zero) represents the equilibrium concentration of silver ions in the solution for the cell reaction

$$Ag^+(aq) + Fe^{2+}(aq) \rightleftharpoons Fe^{3+}(aq) + Ag(s)$$

Since the concentrations of $Fe^{2+}(aq)$ and $Fe^{3+}(aq)$ in the other half-cell are known,

$$K_c = \frac{[Fe^{3+}(aq)]_{eqm}}{[Ag^+(aq)]_{eqm}[Fe^{2+}(aq)]_{eqm}}, \text{ can be calculated.}$$

Note. An alternative to the above procedure would be to alter the $Fe(II)/Fe(III)$ ratio by a continuous variation method, whilst maintaining a fixed silver ion concentration in the other half-cell. This leads to an identical result for K, but the e.m.f. values observed are smaller than those measured in the suggested procedure, thus causing larger experimental errors.

Experiment 17.3c

To find the equilibrium constant for the reaction
$Ag^+(aq) + Fe^{2+}(aq) \rightleftharpoons Fe^{3+}(aq) + Ag(s)$ by titration

Each student or pair of students will need:

0.10M iron(II) nitrate, prepared as described in experiment 17.3b
0.10M silver nitrate

and either

1M sodium chloride
Phosphoric acid
Standardized potassium permanganate solution, approximately 0.02M

or   Standardized potassium thiocyanate solution, approximately 0.1M

Procedure

Reaction mixtures are prepared by adding together 25.0 cm^3 each of the iron(II) nitrate and silver nitrate solutions in dry 100 cm^3 conical flasks. The flasks should be well stoppered and then left to stand for approximately 30 minutes, with occasional shaking.

After equilibrium has been attained, 25.0 cm^3 of the supernatant solution are carefully withdrawn from above the silver precipitate and analysed by one of the following methods, both of which are equally satisfactory.

Method 1 – The following are added to the solution in the order given:

5 cm^3 of 1M NaCl,
10 cm^3 of dilute H_2SO_4, and
$1-2 \text{ cm}^3$ of concentrated H_3PO_4

The mixture is then titrated against the standardized potassium permanganate solution

Method 2 – The 25.0 cm^3 sample of the solution is titrated with the standardized potassium thiocyanate solution, the iron(III) in solution serving as an indicator. The standardization of the thiocyanate may be carried out with the silver nitrate solution used for the experiment. Details of the reactions upon which these two methods depend are given in the *Students' Book*.

Calculation of K_c

The concentrations of silver ion and iron(II) in the equilibrium solution are the same and can be calculated from the titration result. The concentration of iron(III) is calculated by subtracting the concentration of iron(II) remaining in the equilibrated solution from the original concentration of iron(II).

There are appropriate questions and instructions in the *Students' Book* to enable them to do this calculation for themselves.

Comparison of values of K_c obtained by the two experiments

Values of K_c obtained by these methods will probably lie in the range 10 to 18 dm^3 mol^{-1}, but there may well be no discernible difference statistically between results obtained by the first and those obtained by the second methods.

Comparison of 'experimental' values of K_c (from experiments 17.3b and 17.3c) with calculated values (from standard redox potentials)

Students should be asked to calculate the value of K also from the appropriate E values. The calculation is as follows:

$$Fe^{3+}(aq), Fe^{2+}(aq)|Pt \quad E^\ominus = +0.77 \text{ V}$$

$$Ag^+(aq)|Ag(s) \qquad\qquad E^\ominus = +0.80 \text{ V}$$

$$\therefore E^\ominus \text{cell} = -(+0.77)+0.80 \text{ V}$$

$$= +0.03 \text{ V}$$

$$+0.03 = \frac{2.3 \times 8.31 \times 298}{1 \times 96\,500} \lg K_c$$

$$0.505 = \lg K_c$$

$$\therefore 3.2 \text{ dm}^3 \text{ mol}^{-1} = K_c \text{ at } 298 \text{ K}$$

More able students could be asked to speculate about possible reasons for the high experimental values. For example, could this discrepancy be attributed to experimental errors? One factor worth considering is the ease with which iron(II) in the near-neutral solution is oxidized by atmospheric oxygen. Contamination of the iron(II) sulphate used to make the iron(II) nitrate with iron(III) is another.

Activities and activity coefficients

The chief reason for the difference between the experimental and theoretical K values lies in the fact that due to ionic 'congestion' in solution the activities of the ionic species are considerably less than their concentrations. The question of ionic 'congestion' or 'crowding together' has been raised, in passing, in Topic 15 when discussing hydrogen ions in solution. The following discussion in terms of activities and activity coefficients here is not intended for students, although with a group of able students the teacher may consider it useful to discuss it in outline with them.

In electrolyte solutions of finite concentration, electrostatic attraction forces are operative between oppositely charged particles. These forces have the effect of restricting the mobility of the individual particles and hence reduce the 'effectiveness' or 'activity' of the ionic species as far as their chemical behaviour is concerned. To allow for this departure from 'ideal' behaviour, activity coefficients are introduced, defined as the ratio of the effective concentration, or activity, to the stoichiometric concentration of the species, i.e.

$$f_x = \frac{a_x}{[x]}$$

where a_x is the activity of x
$[x]$ is the concentration of x
f_x is the activity coefficient of x

The value of the activity coefficient depends on a number of factors, including ionic charge, ionic size, temperature, and total ionic concentration in solution. Table 17.3h may help to illustrate this.

Ions (in aqueous solution)	Ionic diameter/nm	Total ionic concentration/mol dm^{-3}			
		0.001	0.01	0.1	0.2
		Activity coefficient (f_x)			
H^+	0.9	0.975	0.933	0.860	0.83
K^+, Ag^+, Cl^-, Br^-	0.3	0.975	0.925	0.805	0.76
$Sr^{2+}, Ba^{2+}, SO_4^{2-}$	0.5	0.903	0.744	0.465	0.39
$Ca^{2+}, Fe^{2+}, Zn^{2+}$	0.6	0.905	0.749	0.485	0.42
$Al^{3+}\ Fe^{3+}$	0.9	0.802	0.540	0.245	0.17
$[Fe(CN)_6]^{4-}$	0.5	0.668	0.310	0.048	0.021

Table 17.3h
Approximate individual activity coefficients of ions in water at 25 °C

The equilibrium constant of the $Ag^+(aq)/Fe^{2+}(aq)$ reaction as found by experiments 17.3b and 17.3c was defined in terms of the *concentrations* of the appropriate ions, whereas the 'theoretical' equilibrium constant (obtained from standard redox potentials) is based on effective ionic concentrations or activities. The relationship between K_{theor} and K_{exp} may be deduced as follows:

$$K_{theor} = \frac{a_{Fe^{3+}(aq)_{eqm}}}{a_{Ag^+(aq)_{eqm}} \times a_{Fe^{2+}(aq)_{eqm}}}$$

and

$$K_{exp} = \frac{[Fe^{3+}(aq)]_{eqm}}{[Ag^+(aq)]_{eqm}[Fe^{2+}(aq)]_{eqm}}$$

Since $a_{Fe^{3+}(aq)} = [Fe^{3+}(aq)] \times f_{Fe^{3+}(aq)}$ etc.

it follows that

$$K_{theor} = K_{exp} \times \frac{f_{Fe^{3+}(aq)}}{f_{Ag^+(aq)} \times f_{Fe^{2+}(aq)}}$$

Assuming the total ionic concentrations in experiments 17.3b and 17.3c to have been approximately 0.2M and using the activity coefficients in table 17.3h,

$$\frac{f_{Fe^{3+}(aq)}}{f_{Ag^+(aq)} \times f_{Fe^{2+}(aq)}} = \frac{0.17}{0.76 \times 0.42} \approx 0.5$$

so that

$$K_{theor} = 0.5\, K_{exp}$$

Values of K_c greater even than $6.4 \ dm^3 \ mol^{-1}$ are frequently met with, however, in this experiment, especially if there is iron(III) present in the iron(II) sulphate starting material.

$-\Delta G^\ominus$ as the work of a cell reaction

At the beginning of this section, $\Delta G^\ominus$ was defined as

$$\Delta G^\ominus = -2.3\, RT \lg K$$

(for some particular temperature)

and we have found earlier that

$$E^\ominus_{cell} = \frac{2.3\, RT}{zF} \lg K_c$$

i.e. $zFE^\ominus = 2.3\, RT \lg K_c$

Comparing these two equations

$$\Delta G^\ominus = -zFE^\ominus$$

$\Delta G^{\ominus}$ is an energy and $zFE^{\ominus}$ is a work quantity (in units of joules). The equation $\Delta G^{\ominus} = -zFE^{\ominus}$ therefore gives us a way of picturing $-\Delta G^{\ominus}$ as the *work* we could expect to obtain from a cell if the potential difference across the terminals of the cell were its e.m.f.

This will link up with the present work the experience of those students who have earlier followed the Nuffield Chemistry sample scheme in which $-\Delta G$ was introduced as 'the work a chemical reaction can do' (see *The Sample Scheme Stages I and II, The Basic Course,* Topic 23). It should be remembered, however, that, because in this topic (Topic 17) $-2.3\,RT\lg K$ has been used as a definition, it is $\Delta G^{\ominus}$ and not ΔG which we have considered throughout. Of course, it is still true that

$$\Delta G = -zFE$$

for *any* cell reaction, but it should be made clear to the students that the cell which corresponds to $\Delta G^{\ominus}$ is one with cell electrolytes at unit concentration (strictly speaking unit activity) and gases at one atmosphere pressure.

Experiment 17.3d
The work of a cell reaction
The purpose of this experiment is to enable students to determine the limiting work available from a cell, or a selection of cells, and hence to calculate $\Delta G^{\ominus}$ for the cell reaction.

Each student or pair of students will need
> Daniell cell with amalgamated zinc rod
> Molar solutions of copper(II) sulphate and zinc sulphate
> High resistance voltmeter (0–3 V) or potentiometer
> Milliammeter
> Connecting leads
> Rheostat (up to 1 or 2 kΩ)
> Graph paper

Cell diagram
> $Zn(s)|Zn^{2+}(aq)|Cu^{2+}(aq)|Cu(s)$

> *Alternatively students could be asked to study one of the following cells:*

> $Zn(s)|Zn^{2+}(aq)|Ag^{+}(aq)|Ag(s)$

> $Cu(s)|Cu^{2+}(aq)|Ag^{+}(aq)|Ag(s)$

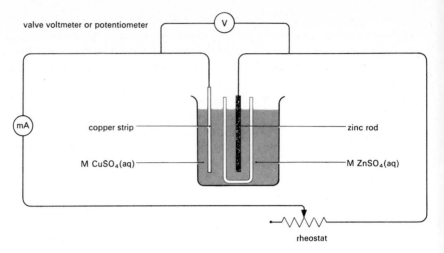

Figure 17.3n

Procedure

The rheostat is set to its lowest setting and the cell voltage and current are noted. The resistance is increased until the p.d. is about 0.1 V more and the new voltage and current noted. This procedure is repeated until the whole resistance of the rheostat is in circuit.

Treatment and discussion of results

A plot of the current flowing in the circuit against the p.d. across the cell terminals can be extrapolated to 'zero current' so giving the e.m.f. $E^\ominus$ of the cell. The work corresponding to this e.m.f. could be calculated (as $zFE^\ominus$). However, in order to let students see that this work is a limiting value, it may be preferable for them to calculate, for each p.d. reading, the work per mole of reaction which the cell could do and then plot these work values against the cell p.d. to obtain the limiting work.

The *Students' Book* contains full details of this treatment and also explains the limiting nature of the work we are discussing in terms of the time it would require for the reaction to take place.

The limiting work, in J mol^{-1}, can be compared with the values of $\Delta G^\ominus_{298}$ listed for some cell reactions in table 4 in the appendix to this topic in the *Students' Book*.

This interpretation of $\Delta G^{\ominus}$ as $-$ (the limiting value of the work obtainable from a cell) enables us to present $\Delta G^{\ominus}$ as a *difference* (equal to $-zFE^{\ominus}$) between energies which may be associated with the products and reactants of the reaction. Such energies are called the standard free energies, $G^{\ominus}$, of the products and reactants, that is

$$\Delta G^{\ominus} = G^{\ominus}\,(\text{products}) - G^{\ominus}\,(\text{reactants})$$

The individual free energies cannot themselves be measured, but only the energy differences (as $-zFE^{\ominus}$). This is a point we shall take up again in the next section (17.4).

Other ways of finding $\Delta G^{\ominus}$

The determination of values of $\Delta G^{\ominus}$ from e.m.f.s affords a useful indirect method of determining K values for reactions between ions in solution, but there remains the question as to how $\Delta G^{\ominus}$ can be determined (independently of the direct measurement of equilibrium constants) for gas reactions.

The *Students' Book* mentions very briefly and in general terms that it is indeed possible to determine $\Delta G^{\ominus}$ independently 'for example by calorimetry' but takes the matter no further.

This is so that only those teachers who wish to pursue the subject of entropy with their students need do so; this development is not intended to be an essential part of this course.

One way in which $\Delta G^{\ominus}$ for gas reactions can be determined, independently of equilibrium constants, is by means of the relation

$$\Delta G^{\ominus} = \Delta H^{\ominus} - T\Delta S^{\ominus}$$

A simple way of deriving this expression is outlined on page 162 of this *Teachers' Guide*, where a comparison is made between $\Delta G^{\ominus}$ and $\Delta H^{\ominus}$ for some reactions. (A simpler derivation in fact stems directly from the definition of G as $H-TS$.)

In this expression $\Delta H^{\ominus}$ can be determined by standard calorimetric procedures (see Topic 7). $\Delta H^{\ominus}$ can also in fact be determined spectroscopically in some cases. $\Delta S^{\ominus}$ can be determined by the use of standard or 'third law' entropies of substances, also determined calorimetrically or from spectroscopy and statistical mechanics as described on pages 172 to 176.

17.4 **Free energy in chemistry**

Objectives

The main aims of this section are:

1   To enable students to use standard free energy data, as published in tables, to calculate equilibrium constants for chemical reactions.

2   To show how free energy data can be interpreted in terms of the relative energetic stabilities of compounds.

Timing

Three or four periods should be sufficient for this section.

Suggested treatment

For this treatment OP transparencies numbers 13, 16, 98, and 99 will be useful.

At the end of section 17.3 in the *Students' Book* it was suggested that standard free energies could be thought of as an energy *difference* between free energy values assignable to the 'products' and 'reactants' of a chemical reaction. This can form a useful starting point for the discussion of standard free energy of formation of compounds with which this section of the *Students' Book* begins.

The treatment can be exactly parallel to that in Topic 7 (Energy changes and bonding) where the standard heat of formation of compounds was introduced. This is in fact a good point for students to revise that part of Topic 7.

The following points (which are treated more fully in the *Students' Book*) could be brought out during the discussion.

1   Just as in the case of enthalpies, or of potential energy, absolute values of free energies are not known, so it is convenient to choose some base-line, or arbitrary zero, from which to measure the standard free energies of substances. The convention chosen for free energies is the same as that for enthalpies, namely that

at 760 mmHg pressure

and 298 K

with the elements in the physical states normal under these conditions, the standard free energies of the elements are zero.

2   The standard free energy of formation of any compound at 298 K, $\Delta G^{\ominus}_{f,298}$, is the standard free energy change when one mole of the compound at 298 K is formed from its elements in physical states normal at 760 mmHg and 298 K

3   It follows necessarily from this convention that $\Delta G^{\ominus}_{f,298}$ [elements in physical states normal at 760 mmHg and 298 K] $= 0$.

4   It is possible therefore, using this convention, to tabulate standard free energies of formation of *compounds* rather than standard free energies relating to specific reactions. (In fact, of course, the standard free energy of a compound does really refer to a reaction, namely the formation of one mole of a compound from its elements in physical states normal at 760 mmHg and 298 K.) The tabulation of standard free energy data relating to compounds rather than reactions, however, makes for very general and flexible use of such tables.

In dealing with ions, the standard free energy of the hydrogen ion is taken as zero, that is

$$\Delta G^{\ominus}_{f,298}[\text{H}^+(\text{aq})] = 0.$$

This corresponds to the convention of regarding the standard potential of the hydrogen electrode as zero volts.

How to calculate equilibrium constants from standard free energies of formation

Examples (1) to (3) in the *Students' Book* are intended to show how the equilibrium constant for a reaction can be calculated from tabulated values of standard free energies of formation. The examples chosen are for reactions involving gases and reactions between ions in solution. The teacher can invent many other examples using the data given in tables 1 to 7 in the appendix to this topic in the *Students' Book*.

Variation of the stabilities of compounds in the Periodic Table

In Topic 7 a careful distinction was made between kinetic and energetic stabilities of compounds. Since $\Delta G^{\ominus}_{f,298} = -2.3\,RT\lg K_f$ where K_f is the equilibrium constant for the reaction in which a compound is formed from its elements under standard conditions, the value of $\Delta G^{\ominus}_{f,298}$ for any compound is an indication of the *energetic* stability of that compound relative to formation from its elements in their standard states. A graph of $\Delta G^{\ominus}_{f,298}$ for a given set of compounds, for example the oxides of the elements, plotted against the atomic numbers of the elements, indicates the variation in energetic stability of the compounds across the Periodic Table.

Figures 17.4a, b, c, and d show the standard free energy of formation at 298 K of various compounds plotted against the atomic numbers of the compound forming elements. The diagrams are also reproduced in the *Students' Book*.

Oxides

For the general reaction between an element M and oxygen, the following equations may be formulated:

1 $\dfrac{x}{y}M(s) + \frac{1}{2}O_2(g) \rightarrow \dfrac{1}{y}M_xO_y(s)$

2 $M(s) + \dfrac{y}{2x}O_2(g) \rightarrow \dfrac{1}{x}M_xO_y(s)$

Both equations allow a comparison to be made between the standard free energies of formation of various metal oxides, the first equation having been written in terms of 1 mole of oxygen atoms as the 'common unit', the second one referring to 1 mole of M atoms instead. The free energy spectrum, based on equation (2) for oxides* is given in figure 17.4a.

Chlorides

Figure 17.4b is a similar diagram for chlorides. The general equation here is

$$M(s) + \frac{x}{2}Cl_2(g) \rightarrow MCl_x(s)$$

Not much time need be spent on a consideration in detail of these diagrams. It is easily seen that for these groups of compounds there is a systematic variation in the free energy of formation across the Periodic Table; stabilities increase and decrease in accordance with the periods.

Students may be interested in looking more closely at some of the free energy relationships across a transition metal series. It is suggested that their attention be drawn to the stability of the various transition metal oxides and the free energy of formation of the metal ions having different oxidation numbers.

Another possible study is that of the free energies of formation of hydrides and comparison with $\Delta H^\ominus_{f,298}$ data for hydrides.

The standard free energies of formation of oxides and chlorides are plotted against the atomic number of the element forming them in the *Students' Book* as figures 17.4a and 17.4b respectively. The values of $\Delta G^\ominus_{f,298}$ for the oxides and hydrides are given in tabular form as tables 5 and 6 in the appendix to this topic in the *Students' Book*, and $\Delta H^\ominus_{f,298}$ for the hydrides as table 7. Values of $\Delta G^\ominus_{f,298}$ for metal ions are given in the *Book of Data*.

*Where metals have various oxidation numbers in forming oxides, the data given are those for the most common oxides.

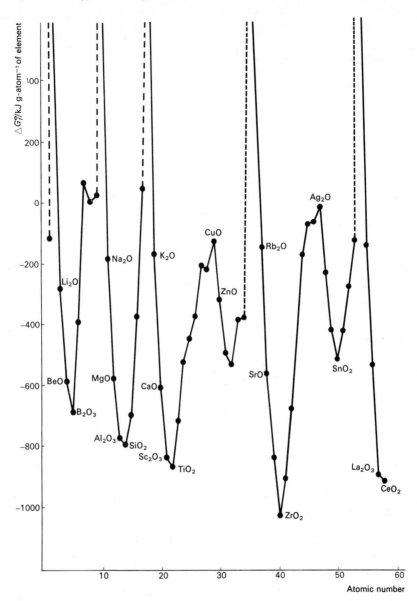

Figure 17.4a
Free energy spectrum of oxides at 298 K.
[General reaction: $M(s) + \frac{y}{2x}O_2(g) \rightarrow \frac{1}{x}M_xO_y(s)$]

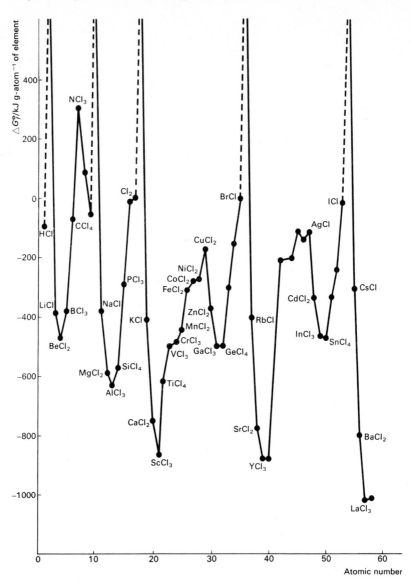

Figure 17.4b
Free energy spectrum of chlorides at 298 K.
[General reaction: $M(s) + \frac{x}{2}Cl_2(g) \rightarrow MCl_x(s)$]

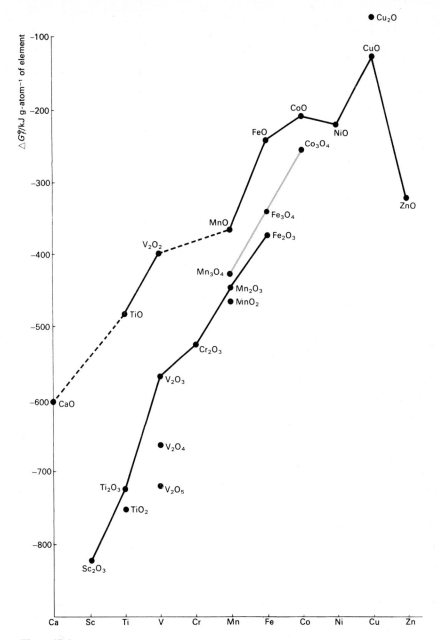

Figure 17.4c
Free energy diagram for metal oxides of first transition series.
[General reaction: $M(s) + \frac{y}{2x}O_2(g) \rightarrow \frac{1}{x}M_xO_y(s)$]

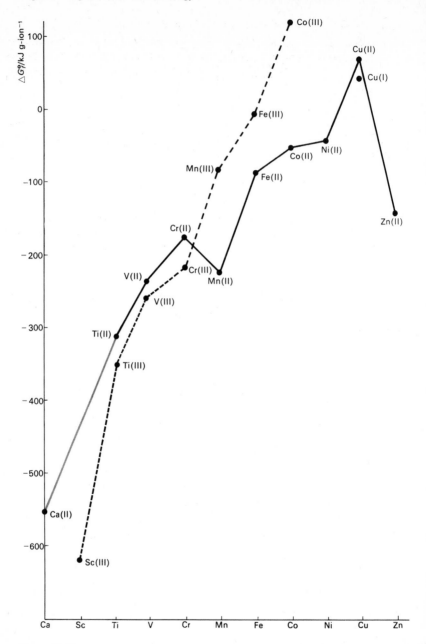

Figure 17.4d
Free energies of formation of ions in aqueous solution for first transition metal series.

The extraction of metals – Ellingham diagrams

Ellingham diagrams are an effective way of examining the standard free energy changes, at various temperatures, associated with the reactions between various elements and oxygen. Such diagrams are also used for sulphides and chlorides as well as oxides.

A sample treatment, in some detail, is in the *Students' Book*. It is along the lines of that in the Nuffield Chemistry *Sample Scheme Stage III*, pages 70 to 73. See also *Principles of extraction of metals* by D. J. G. Ives, R.I.C. Monographs for Teachers, No. 3, page 17.

Kinetic and energetic stabilities

The two examples in this sub-section in the *Students' Book* are intended as a reminder once more of the point made in Topic 7 and in section 17.1 of the present topic that kinetic factors can often prevent the attainment of equilibrium.

Energetics in life processes

The material in this final sub-section in the *Students' Book* is an attempt to show how energy calculations can be applied to problems in related fields, in this case the problem of photosynthesis

Summary

At the end of this topic students should

1   Understand that the fundamental tendency in any chemical process is the attainment of equilibrium.

2   Understand the relationship between the equilibrium constant and $\Delta H^{\ominus}$ and be able to find equilibrium constants for the same (gas) reaction at different temperatures.

3   Understand the relationship between the equilibrium constant and $\Delta G^{\ominus}$ and be able to find equilibrium constants indirectly for both reactions between gases and reactions between ions in solution.

4   Be able to use tables of the standard free energy of formation of compounds as a means of calculating equilibrium constants indirectly.

Suggested further reading (for the teacher)

*Bent, H. A. (1965) *The second law*, Oxford University Press, New York.

*Nuffield Advanced Physics, Teachers' Guide to 'Change and chance' (in preparation).

*Millen, D. J., ed. (1969) *Chemical energetics and the curriculum*, Collins.

Caldin, E. F. (1958) *An introduction to chemical thermodynamics*, Oxford University Press.

Denbigh, K. (1966) *The principles of chemical equilibrium*, 2nd edition, Cambridge University Press.

Rushbrooke, G. S. (1949) *Introduction to statistical mechanics*, Cambridge University Press.

Kauzmann, W. (1967) *Thermodynamics and statistics*, Benjamin.

The ICI Film series 'The Laws of Disorder' (Millbank Films) may also be found useful for both teachers and students.

* Specially recommended.

Answers to problems in the *Students' Book*

(A suggested mark allocation is given in brackets after each answer.)

1

i 1.78×10^4 atm   (1)

ii 2.00×10^6 atm   (3)

iii $\Delta H^{\ominus}$ is independent of T in the range 600–1000 K   (1)

2

i (a) $+2.32$ (b) 2.09×10^2 atm   (2)

ii 1.23×10^3 atm   (2)

iii 1500 K   (1)

3

i -65.4 kJ   (2)

ii 1.35 dm^3 mol^{-1} (3)

4

i From gradient of graph $\Delta H^{\ominus} = +59.8$ kJ   (2)

ii $\Delta G^{\ominus}_{423} = -23$ kJ   (3)

5

i $Pt|I^-(aq), I_2(aq)| Br_2(aq), Br^-(aq)|Pt$ (1)

ii $+0.55$ V   (1)

iii 2×10^9 (for equation as written in the question)   (3)

6

i $MnO_4^-(aq) + 8H^+(aq) + 5e^- \rightleftharpoons Mn^{2+}(aq) + 4H_2O(l)$ (2)

   and $5Fe^{3+}(aq) + 5e^- \rightleftharpoons 5Fe^{2+}(aq)$ (2)

ii $Pt|Fe^{2+}(aq), Fe^{3+}(aq)|[MnO_4^-(aq) + 8H^+(aq)],$

$[Mn^{2+}(aq) + 4H_2O(l)]|Pt$ (2)

iii $+0.74$ V   (2)

iv $K_c = 10^{125}$ (2)

7

i -50.8 kJ   (1)

ii Yes   (1)

iii No   (1)

iv Answer should refer to distinction between kinetic and energetic
stability.   (2)

8

i Both iron(II) oxide and iron(III) oxide are stable relative to formation
   from the elements
   $$\Delta G^{\ominus}_{f,298}[\text{FeO}] = -244.3 \text{ kJ mol}^{-1}$$
   $$\Delta G^{\ominus}_{f,298}[\text{Fe}_2\text{O}_3] = -741.0 \text{ kJ mol}^{-1} \quad\quad\quad (2)$$

ii Iron(II) oxide is unstable relative to oxidation by the air to iron(III)
   oxide
   $$(2\text{FeO(s)} + \tfrac{1}{2}\text{O}_2\text{(g)} \rightarrow \text{Fe}_2\text{O}_3\text{(s)}; \Delta G^{\ominus}_{298} = -252 \text{ kJ}) \quad\quad\quad (3)$$

9

i -147.4 kJ   (2)
ii Greater   (2)
iii $K_c = 4.5 \times 10^{12} \text{ dm}^6 \text{ mol}^{-2}$ (2)
iv 4 marks for reasonable suggestions   (4)

10

i -9 kJ   (2)
ii 38.2   (2)
iii Yes   (1)

Topic 18
Carbon compounds with large molecules

Objectives

The aims of this topic can be summarized as follows:

1 By showing the structural formulae of a range of naturally occurring organic macromolecules, to let the students become aware that the structures of such compounds can be extremely complex, and although created easily within the living organism, present the most challenging of structural and synthetic problems yet met by chemists.

2 By consideration of soap and non-soapy detergents, to show how knowledge of the relationship of structure and properties can lead to man's imitation of nature, by the creation of similar but not identical materials having their own special features.

3 By focusing attention on one class of naturally occurring macromolecules, the proteins, to show what can be learnt about the structure of large molecules, and to investigate this experimentally.

4 By consideration of some polymerization reactions, to show the success man has had in building-up large molecules from small ones, and to investigate experimentally how the properties of materials having large molecules depend upon their molecular structure.

Contents

Introduction.

18.1 Detergents.
18.2 The nature of proteins.
18.3 The chemical investigation of proteins.
18.4 The structural investigation of proteins.
18.5 Enzymes.
18.6 An examination of some plastics.
18.7 The laboratory preparation of some plastics.
18.8 The invention and discovery of synthetic polymers.

Timing

About four weeks.

Introduction

This topic is built on a number of ideas students have already met and therefore affords an opportunity for revision and application of those ideas, namely

X-ray diffraction (structure of proteins and plastics)
bonding (structure of the peptide group)
intermolecular forces (action of detergents, structure of proteins)
solvation (action of detergents)
carbon chemistry (amino acid chemistry, polymerization reactions).

After a brief introduction to naturally occurring macromolecules, the topic deals first with detergents as these are relatively simple molecules with which students are likely to have some familiarity. The objective is to indicate how the study of natural materials can lead to the development of alternative synthetic materials.

The topic continues with an account of proteins and considers how their structure is elucidated. There is a section on enzymes and a brief indication of their use as industrial catalysts.

The last theme of the topic is plastics, originally substitute materials but now valuable materials in their own right. The preparation, structure, and practical properties of plastics are all considered.

The topic is developed in the following way:

The idea of macromolecules
↓
lipids, proteins, sugars and nucleic acids
↓   ↓
soap   amino acids
↓
detergents
↓
The idea of synthetic materials → plastics → monomers → polymers

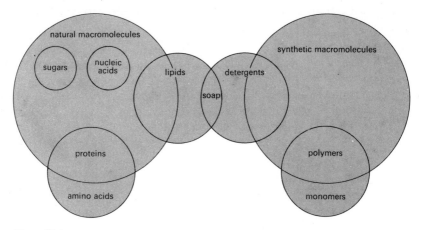

Figure 18.A

18.1 Detergents: how a problem can be solved by a natural product and by a synthetic product

Objectives

1 To establish the molecular structure necessary for a compound to have detergent properties.

2 To show how this knowledge has resulted in detergents from natural sources (soap) being replaced by synthetic detergents.

3 To indicate briefly that the use of synthetic detergents has disadvantages, but these are different from those associated with soap.

Timing

About five periods.

Suggested treatment

For this treatment OP transparencies numbers 104 and 122 will be useful.

Students should be familiar with the background to this section from their O-level course, and be aware of the existence of hard and soft waters and the relative behaviour of soap and synthetic detergent. Teachers may therefore feel that a discussion of detergent function coupled with experiments 18.1c and 18.1d will be an adequate treatment of this section.

Confusion may arise in this section unless students are careful to note the correct usage of various terms:

detergents are all materials used for cleaning.

soapy detergents are commonly known as 'soap'.

soapless detergents are commonly known as 'synthetic detergents', or even just 'detergents'.

$$\textit{sulphation} \text{ is the introduction of the } -O-\overset{\overset{\displaystyle O}{\|}}{\underset{\underset{\displaystyle O}{\|}}{S}}-O^- \text{ group into a molecule.}$$

$$\textit{sulphonation} \text{ is the introduction of the } -\overset{\overset{\displaystyle O}{\|}}{\underset{\underset{\displaystyle O}{\|}}{S}}-O^- \text{ group into a molecule.}$$

In the *Students' Book* after a statement of the composition of natural fats and oils students are given instructions for the preparation of two detergents from castor oil.

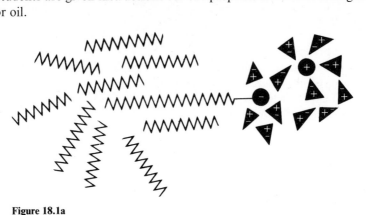

Figure 18.1a

Representation of a detergent molecule with an ionic group in water and a hydrocarbon chain in grease.

Students' attention is next drawn to the need to produce synthetic detergents. Very large quantities of detergents are now used and natural oils, which are more important as foodstuffs anyhow, can no longer meet the demand. By-products of the petroleum industry must therefore be used, but to select the best starting materials the relationship between detergent properties and molecular structure must be further investigated. To illustrate this point experiment 18.1c investigates the influence on the surface tension of water of chain length in sodium salts of carboxylic acids, $CH_3(CH_2)_nCO_2^-Na^+$.

In experiment 18.1d the preparation of a synthetic soapless detergent from dodecanol and chlorosulphonic acid is described. The product can be tested against soft and hard water so that students can be reminded of the relative merits of soapy and soapless detergents in hard water districts.

They are then asked to consider how detergents work. From their knowledge of Topic 10 (Intermolecular forces) students should be able to propose that a detergent must have a long hydrocarbon chain which will dissolve in grease, using van der Waals' forces of attraction, and must also have an ionic group which can be solvated, and hence will tend to remain in contact with water.

As the ionic group of a soap is a carboxyl group, hydrogen bonding will be the major force of attraction between the detergent and water molecules.

Some types of dirt particles such as carbon are removed in the cleaning process in a similar way to grease but others are removed by a slightly different process:

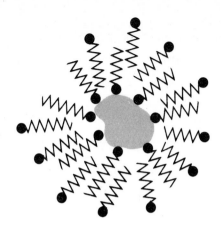

Figure 18.1b
A representation of a rust particle with a soluble layer of adsorbed detergent molecules.

particles such as rust attract the ionic *head* of the soap particle and then a second layer is built up.

A final section describes some of the varied detergents now available and mentions problems associated with their use, for example foam at sewage works. The Nuffield film loop 'Problems in the use of detergents' could be shown again here.

For homework students could read one of the background booklets available (see 'Supporting material' at the end of this section).

Experiment 18.1a
Preparation of a soapy detergent

Each student or pair of students will need:
  100 cm³ beaker
  Tripod and gauze
  Sodium hydroxide, solid
  Castor oil
  Sodium chloride, solid
  M calcium chloride solution
  Apparatus for suction filtration
  Test-tube
  Conical flask

Procedure
Full details are given in the *Students' Book*. The product of sodium ricinoleate should lather well and give precipitates with calcium salts and hydrochloric acid.

Experiment 18.1b
Preparation of Turkey Red oil, a soapless detergent

Each student or pair of students will need:

Test-tubes
Castor oil
Concentrated sulphuric acid
Dropping tube
Conical flask

Procedure

The castor oil is sulphated in the reaction but more complex changes also occur. The product is an impure mixture and this experiment should be treated as a short ten-minute exercise.

Experiment 18.1c
To investigate the action of various sodium salts of carboxylic acids on the surface tension of water

Each student or pair of students will need:

6 test-tubes
6 capillary tubes of uniform bore (the length will depend on the diameter of the tubes available: they must be long enough to accommodate the capillary rise of water which should be about 6 cm).
Ruler for measuring capillary rise

access to beakers of 0.01 M *solutions of:*

Sodium acetate
Sodium butanoate
Sodium hexanoate
Sodium octanoate
Sodium decanoate
Sodium dodecanoate

Procedure

For satisfactory results it is essential that capillary tubes are cut from the same length of thick-walled capillary tubing to be uniform bore, and are well cleaned before use. They should be rinsed in sodium hydroxide solution, water, and acetone, and then dried.

Some of the sodium salt solutions may have to be made by dissolving the appropriate amount of free acid in 0.01M sodium hydroxide.

The procedure is described in the *Students' Book*. Students plot a graph of capillary rise against number of carbon atoms in the salt. Sodium dodecanoate depresses the surface tension by about 30 per cent. It would be expected that sodium hexadecanoate (16 carbon atoms) would depress it by 70 per cent.

Experiment 18.1d
Preparation of a synthetic soapless detergent

Each student or pair of students will need:

Two 100 cm^3 beakers
Cold water bath
Thermometer, 0–110 °C
Two *dry* 10 cm^3 measuring cylinders
Dodecanol
Chlorosulphonic acid
Triethanolamine (optional) or 2M sodium hydroxide solution
Sodium bicarbonate

Procedure

Full details are given in the *Students' Book*. This experiment must be carried out in a fume cupboard and because of the corrosive nature of chlorosulphonic acid, teachers may prefer to conduct the experiment as a demonstration.

The liquid detergent product is a common component of hair shampoos. This information is not given in the *Students' Book* as the practical application of their crude products would not be wise.

Supporting homework

Reading one of the background booklets available.

Supporting material

Unilever Education Booklets: Revised Ordinary Series No. 1 *Detergents* (1967).
Unilever Education Booklets: Advanced Series No. 4 *The chemistry of glycerides* (1965).
Unilever Laboratory Experiments: *The Preparation of Soap.*
Unilever Laboratory Experiments: *The Preparation of Sodium Alkylbenzene Sulphonate.*
Nuffield film loop 'Problems in the use of detergents'.
16 mm films: 'Chemistry of soapless detergents', Unilever Film Library, Unilever House, Blackfriars, London EC4.
'Detergents up to date', Petroleum Film Bureau, 4 Brook Street, London W1.
Nuffield Chemistry Background Book: *Detergents.*

Summary

As a result of their work in this section students should be aware of the chemical composition of fats. They should know the molecular structures associated with detergent properties and be able to discuss the intermolecular forces involved in detergent action relative to grease. They should also appreciate the comparative behaviour of soapy and synthetic soapless detergents.

18.2 **The nature of proteins**

Objectives
1   To indicate the existence in nature of a class of compounds which we can call 'proteins'.
2   To discuss the amino acid composition of proteins.

Timing
Two periods, one for discussion and one for practical work.

Suggested treatment
For this treatment OP transparency number 56 will be useful.

Students should be tested on their understanding of section 13.4, 'Two different functional groups in the same molecule', before beginning this work, and if necessary asked to revise the material for homework.

The experimental work introduces students to proteins and their associated amino acids. It is useful at this point to introduce materials which may reappear in subsequent sections. Thus commercial meat extracts, 'Oxo' cubes, are labelled as containing monosodium glutamate and the glutamic acid can be subsequently identified by chromatography.

The testing of insoluble material (hair, nail parings, porcupine quill) could be considered. Refluxing overnight with 6M hydrochloric acid will produce the free amino acids, and a residue suitable for paper chromatography can be obtained by evaporation to dryness on a water bath followed by storage in a soda-lime desiccator. Porcupine quill (a fisherman's float or from a zoo) is a valuable example, much discussed later in the topic, and its hydrolysis is rapid.

The relationship, milk → casein (protein extract of milk) → casein hydrolysate (partially hydrolysed protein extract of milk), should be stressed and hence the relationship of protein to amino acids.

Discussion of proteins can be general with the objective of establishing an appreciation of their importance rather than specific facts about specific proteins. Biologists should be able to make a valuable contribution to such a discussion, alternatively the teacher will find additional information in the Special study *Biochemistry*.

The Nuffield Chemistry film loop 'Giant molecules – proteins' could be shown to emphasize the structural relationship of amino acids to protein.

Experiment 18.2
Investigation of protein materials

Each student or pair of students will need access to:

> Protein materials, for example milk, an egg, casein, albumen, pepsin, trypsin, gelatin, glycine, casein hydrolysate, meat extract, porcupine quill.
> Amino acid solutions (see experiment 18.3a)
> 0.02M ninhydrin in acetone (store in refrigerator)
> 2M sodium hydroxide
> 0.1M copper(II) sulphate
> Capillary tubes (organic melting point tubes)
> An oven at 110 °C

Procedure
Full details are given in the *Students' Book*.

As this experiment is open-ended in terms of materials available for testing it is suggested that the experimental work is restricted to a single period and if wished the results obtained by a cooperative effort.

The colours obtained when testing with ninhydrin vary somewhat depending on the amino acids present. The structure and reactions of ninhydrin are not required knowledge but are given in the *Students' Book* as figure 18.2a.

Supporting homework
Revision of section 13.4.
Reading the background booklets listed below.
Collection of protein material for laboratory testing.

Supporting material
Unilever Educational Booklet: Advanced Series No. 3 *The chemistry of proteins*.
Fruton, J. S. 'Proteins' *Scientific American* offprint No. 10.
Kopple, K. D. (1966) *Peptides and amino acids* Benjamin.
Nuffield Chemistry film loop 3–4 'Giant molecules – proteins'.

Summary
Students should now have an elementary knowledge of protein materials and know they consist of amino acids joined by peptide groups.

18.3   **The chemical investigation of proteins**

Objective
To investigate the amino acid composition of protein material by simplified experimental procedures, and discuss the more elaborate procedures required to find the amino acid sequence in a protein chain.

Timing
Four periods, including a double period, will be needed.

Suggested treatment
This section is concerned with the determination of the amino acid sequence in protein molecules, and the account in the *Students' Book* is based on Sanger's work on insulin. Fuller accounts of this work are available in the two *Scientific American* offprints listed under supporting material.

The chromatographic separation suggested as experiment 18.3a does no more than indicate the potentialities of the method although a skilled student with extra time could attempt a two-way chromatogram. Experiment 18.3a requires a full double period so the procedure and apparatus are best discussed before-hand. During the experiment there is a fifty minute interval while the chromatogram is developed. This time could be used for discussion of Sanger's work on insulin. The Nuffield film loops 'Applications of paper chromatography' and 'Two-way paper chromatography' can also be used.

To start the discussion it might be fruitful for students to do the final exercise in the *Students' Book* of cutting up a 7 unit chain (leaving the 10 unit chain for homework) and attempting to reassemble the chain.

For the various questions posed in the text students should be able to produce answers along the following lines:
  1   The hydrolysis of peptide groups works as the reverse reaction to their formation, that is, addition of water to break up the group; elimination of water to form the group.
  2   Examination of the two way chromatogram (figure 18.3b of the *Students' Book*) reveals that the basic compound lysine was more soluble in phenol, the acidic compound glutamic acid was more soluble in the stationary water, and neutral compounds glycine and alanine were intermediate in behaviour.
  3   Sanger found two terminal amino groups on insulin which suggests that it consists of two polypeptide chains cross linked in some way. In insulin two cystine residues act as bridges between the chains:

$$
\begin{array}{ccc}
| & & | \\
NH & & NH \\
| & & | \\
CH-CH_2-S-S-CH_2-CH \\
| & & | \\
CO & & CO \\
| & & |
\end{array}
$$

An important step in Sanger's work was the breaking of these sulphur bridges by oxidation using formic acid and hydrogen peroxide (performic acid), and subsequent separation of the two chains, thereby halving the complexity of the work.

An acidic and a basic amino acid could also form a bridge between protein chains, for example lysine and glutamic acid:

$$
\begin{array}{ll}
| & | \\
NH & NH \\
| & | \\
CH\!-\!(CH_2)_4\!-\!NH_3^+\ ^-O_2C\!-\!(CH_2)_2\!-\!CH \\
| & | \\
CO & CO \\
| & |
\end{array}
$$

4   The action of carboxypeptidase on the protein somatotropin produces 4.0×10^{-5} mole phenylalanine per $\dfrac{1}{45\ 000}$ mole of protein, that is 2 mole of phenylalanine per mole of somatotropin growth hormone, suggesting either one peptide chain with the sequence Phe, Phe, Ala, Leu . . . or two cross-linked peptide chains beginning Phe . . . . J. Ieuan Harris, the author of the original paper quoted, is the major worker in this field.

Experiment 18.3a
Chromatographic separation of amino acids

Each student or pair of students will need:

    1 piece of Whatman chromatography paper CRL/1 and 2 paper clips
    1 dm^3 (polypropylene) beaker and cap, or any glass container of suitable size.

access to:

    0.1 per cent (approximately 0.01M) amino acid solutions in a mixture of water (9 parts), propan-2-ol (1 part)
    Casein hydrolysate or porcupine quill hydrolysate (see 18.2)
    Capillary melting point tubes
    Fresh solvent mixture of butan-1-ol (5 parts), glacial acetic acid (1 part), water (4 parts); 20 cm^3 of upper organic phase for each student
    Spray bottle containing 0.02M ninhydrin in acetone (store in a refrigerator) or ninhydrin aerosol spray
    Spray bottle containing a mixture of methanol (95 parts), M aqueous copper(II) nitrate (5 parts), and 2M nitric acid (a few drops)
    0.880 ammonia in an evaporating basin (fume cupboard)
    An oven at 110 °C

Procedure

Necessary details are given in the *Students' Book*; in order to complete the experiment in a double period the development of the chromatograms is not carried very far. A longer procedure is described in the Special study *Biochemistry*.

The selection of amino acids is unimportant; the six cheapest will do and 10 cm^3 of solution (roughly 0.01 g solid) will last several years. The BDH Amino Acid Reference Collection, price £3.75 is a convenient collection of twenty-four chromatographically homogeneous amino acids. Or for about £2 it is possible to buy cystine, glutamic acid, glycine, tyrosine, and valine.

As a mixture for comparison with known amino acids, casein hydrolysate or porcupine quill hydrolysate are suitable.

The chromatography paper is spotted using melting point tubes. Two or three should be placed in each stock bottle of amino acid solution. The development and spraying of the chromatograms is best conducted in fume cupboards as the combined odours can be unpleasant.

The finished work can be kept in polythene bags.

Notes
1   Cheap multi-pour polypropylene beakers with cardboard caps can be bought from Arnold R. Horwell, Limited, 2 Grangeway, Kilburn High Road, London NW6.
2   Whatman chromatography paper CRL/1 is sold in boxes of 100 sheets.
3   Tall-form plastic spray bottles from Boots are satisfactory if only one-quarter full; otherwise ninhydrin is available in aerosol cans.

Experiment 18.3b
Amino acid sequence in a dipeptide
This experiment is included in the *Students' Book* so they can at least read the experimental details for determining the N-terminal amino acid in a protein. No time is allowed in the course for carrying out this experiment but an enthusiastic student might be able to do so. About two double periods are required.

Each student or pair of students will need:
Glycyl-DL-leucine *or* DL-leucyl-glycine, 10 mg
1-Fluoro-2,4-dinitrobenzene, 2 per cent (v/v) in ethanol, 1 cm^3
0.4M sodium hydrogen carbonate
Butan-1-ol
Graduated dropping tubes, 1 cm^3 and 2 cm^3
Whatman CRL/1 chromatography paper
1 dm^3 beaker or suitable container for chromatography
Capillary melting point tubes
Disposable polythene gloves

Procedure

Necessary details are given in the *Students' Book*.

The dipeptides cost £2 per gramme (from Koch-Light Limited, Colnbrook, Bucks.) but only 10 mg is required for each experiment.

The chromatographic procedure is as experiment 18.3a, except that the chromatograms must be run in the dark.

The result with glycyl-DL-leucine, $NH_2CHCONHCHCO_2H$

$$\underset{\displaystyle H}{|} \qquad \underset{\displaystyle \underset{\displaystyle CH(CH_3)_2}{\underset{\displaystyle |}{CH_2}}}{|}$$

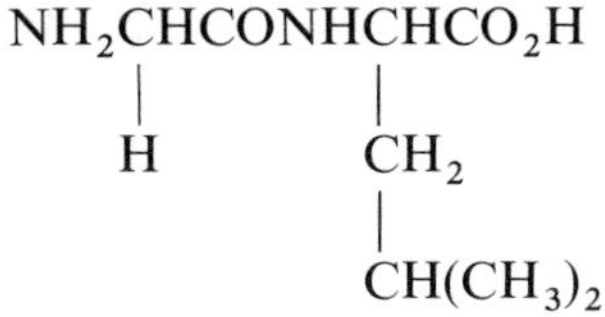

is the formation of DNP glycine

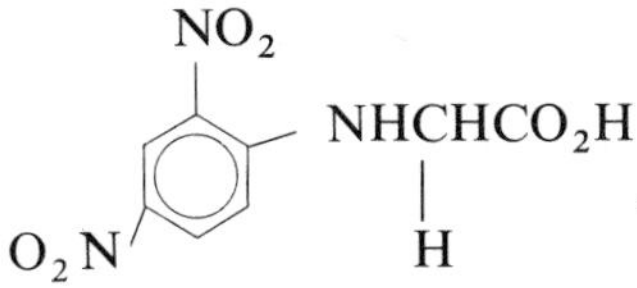

$, R_f\ 0.30-0.35.$

Supporting homework

Carrying out the 7 and 10 chain polypeptide exercises.
Answering the questions in the text of the *Students' Book*.
Reading supporting material.

Supporting material

Thompson, E. O. P. 'The insulin molecule' *Scientific American* offprint No. 42.
Stein, W. H. and Moore, S. 'The structure of proteins' *Scientific American* offprint No. 80.
Nuffield film loops 'Applications of paper chromatography' and 'Two-way paper chromatography'.
16 mm film: 'The structure of proteins', Unilever Film Library, Unilever House, Blackfriars, London EC4.

Summary

Students should now know in principle the method of determining the amino acid sequence of a protein, and have developed some skill in chromatographic technique.

18.4 **The structural investigation of proteins**

Objectives

1   To revise students' understanding of X-ray diffraction, electron density maps, and hydrogen bonding.

2   To review the experimental evidence for the α-helix by a quantitative study of diagrams provided.

Timing

Three periods plus homework.

Suggested treatment

For this treatment the following OP transparencies will be useful: numbers 30, 35, 36, 57–60, and 63.

Many students will find this section of the course difficult; if the teacher considers this will be so for his students, the best treatment would be to concentrate on the revision of related material and only attempt in detail experiment 18.4b on the α-helix.

A fuller discussion of protein structure is contained in the *Scientific American* offprint quoted under 'Supporting material'; this personal account of the work of Pauling might also be of interest to students.

Before starting this section the material in Topic 8 on X-ray diffraction and in Topic 10 on the hydrogen bond could usefully be revised for homework.

The *Students' Book* first attempts to give students a sense of the number of possible conformations for a polypeptide chain by asking them to cut a piece of string to correspond to the length of an insulin molecule on the scale of 0.1 nm = 1 cm. The piece of string has to be 215 cm long.

The 'experiments' in this section do not require a laboratory as they involve the examination of other people's experimental results. Thus experiment 18.4a involves the examination of electron density maps of N-acetylglycine and the cyclic anhydride of glycine (diketopiperazine), so that students will have an understanding of some of the essential parameters of the peptide group and the hydrogen bond in proteins. The results of measurements on acetylglycine and other amino acids are given in table 18.4a. The measured hydrogen bond length in the cyclic anhydride of glycine is 0.29 nm.

	p (nm)	o (nm)	d (nm)	e (nm)	1 (°)	2 (°)	3 (°)	4 (°)
N-acetylglycine	0.132	0.124	0.150	0.145	121	118	121	120
Glycyltyrosine, HCl, H_2O	0.135	0.116	0.153	0.147	117	111	132	116
β-Glycylglycine	0.129	0.123	0.153	0.148	121	114	125	122
Cysteylglycine, $\frac{1}{2}$NaI	0.132	0.121	0.154	0.133	109	126	107	139
Diketopiperazine	0.133	0.125	0.147	0.141	—	—	—	—

Table 18.4a
Comparison of bond lengths and bond angles about the peptide link of various compounds

In discussing the bonding in the peptide group students should be able to recall that a carbon-carbon double bond holds atoms co-planar and without free rotation. A partial double bond in the peptide group can be proposed

and this is confirmed by the shortening of the bond compared to its usual single bond length (C—N from 0.147 nm to 0.132 nm). It might also be noted that the polypeptide chain in naturally occurring materials has a *trans* configuration about the C═N bond.

X-ray fibre photographs are described next in the *Students' Book*. Nuffield diffraction grids are introduced only to illustrate the discussion and their use does not 'prove' anything. The grid with the rectangular pattern of dots should be held close to the eyes and used to view a pearl light bulb about three metres away. In this situation the individual diffraction spots merge into horizontal layer lines (see figure 18.4).

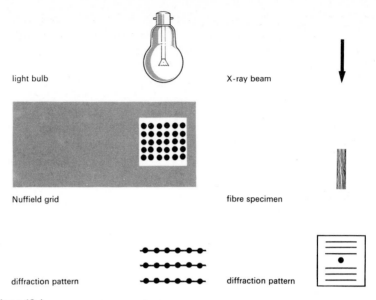

light bulb

X-ray beam

Nuffield grid

fibre specimen

diffraction pattern

diffraction pattern

Figure 18.4
Comparison of optical and X-ray diffraction patterns.

If students can see no connection between a regular pattern of dots and a possible arrangement of polypeptide chains they should look again at figure 10.2g showing the structure of silk where they will see that the nitrogen atoms, for example, form a regular pattern.

As described in the *Students' Book*, synthetic polypeptides were used in the further investigations of the structure of fibrous proteins. Results that were obtained showed that two sets of regular repeats occur, at 0.54 nm and 0.15 nm intervals respectively. The problem was to devise an arrangement for peptide groups that possesses both these repeats. The successful answer, proposed by Pauling and Corey, is the α-helix.

In experiment 18.4b students are asked to look in detail at a drawing, and build a model if possible, of the α-helix so that they explore the structure properly.

The information on naturally occurring materials which concludes the section can be read for interest by the students but detailed discussion in the classroom is not necessary.

It has been proposed that the two varieties of silk are packed with the fibres back-to-back and all their side chains are to one side in the case of Bombyx mori silk, while in the case of Tussore silk the side chains are on both sides of the fibre axis.

Experiment 18.4a
The structure of the peptide group

Each student or pair of students will need:

      Electron density map of acetylglycine
      Electron density map of the cyclic anhydride of glycine
      Protractor
      Ruler

Procedure

The electron density maps are given as illustrations in the *Students' Book* but the teacher may prefer to duplicate copies from a spirit master prepared from the large scale overhead projector printed originals.

The results of the measurements required are listed in table 18.4a and their interpretation is discussed above.

Experiment 18.4b
The α-helix

Each student or pair of students will need:

      Diagram of the α-helix
      Set square (or protractor)
      Ruler
      Ball-and-spoke atomic models

Procedure

The α-helix is illustrated in the *Students' Book* but the teacher may feel that the best procedure is to duplicate copies of the large scale overhead projector printed original.

The questions in the *Students' Book* on the α-helix can be answered as follows:

    1   To trace the hydrogen bonding sequence, number the nitrogen atoms up the helix when it will be found that hydrogen bonds interlink the peptide groups with nitrogen atoms 1—4—7—, 2—5—8—, and 3—6—9—. The sequence is emphasized by colouring the diagram.

    4   In the overhead projector printed original the vertical spacing between nitrogen atoms is 15 mm, the hydrogen bonds are 28 mm long, and the pitch of the helix is 55 mm. These dimensions are in the ratio 1/1.9/3.7 compared to the ratio 1/1.9/3.6 for the molecular dimensions of 0.15 nm, 0.28 nm, and 0.54 nm.

5   To build a ball-and-spoke model of an α-helix the following procedure can be followed.

a   Build a chain of consecutive C=C and C units, using about ten pairs in all and beginning and ending with a C unit.

b   Make the chain *trans* across the double bonds and lay it out straight.

c   Starting from the left, label the first atom in each double bond as the peptide N atom.

d   Orientate the first N—H bond in the chain to point upwards, then join it to the O atom attached to the second (C—O) atom of the fourth double bond using an appropriate length of rubber tubing to represent the extra long hydrogen bond. Make sure the chain spirals up clockwise.

e   Join the second double bond to the fifth and so on up the chain. Alternatively a model of the α-helix is available (see Appendix 3).

Supporting material
Pauling, L. B., Corey, R. B., and Hayward, R. 'The structure of protein molecules', *Scientific American* offprint No. 31.

18.5   Enzymes

Objectives
1   To establish that the class of proteins known as enzymes function as biochemical catalysts.
2   To examine experimentally the factors that influence the catalytic behaviour of enzymes.

Timing
About four periods (including one for discussion).

Suggested treatment
For this treatment OP transparencies numbers 61 and 62 will be useful.

Students will probably only have time for two of the three experiments. It is suggested that they carry out either the saliva or the milk investigation first and finish with the urease experiment which is quite quick. The urease experiment is more fully investigated in the *Biochemistry* Special study.

The high temperature failure of enzymes is ascribed to a 'denaturing' process common to proteins in which their hydrogen-bonded structure breaks down and the enzyme loses its particular shape.

The influence of pH might be illustrated by reference to lysozyme whose function depends on a particular amino acid having an ionized carboxyl group. The account of the work on lysozyme is intended as background reading to the topic and a detailed understanding is not expected. The discussion period is to establish the points made by the practical work – the milk and urease experiments would be suitable on a larger scale as teacher demonstrations.

Experiment 18.5a
The specificity of urease
Each student or pair of students will need:

1 per cent solution (cloudy suspension) of urease active meal (from jack beans), 5 cm^3
0.25M urea, 5 cm^3
0.25M thiourea, 5 cm^3
0.25M methylurea, 5 cm^3
0.25M acetamide, 5 cm^3
Other similar solutions, 5 cm^3
0.01M hydrochloric acid, 10 cm^3
Universal indicator solution
Two dropping tubes
Stopclock

Experiment 18.5b
The activity of α-amylase
Each student or pair of students will need:

(Saliva)
1 per cent fresh starch solution, 70 cm^3
0.1M hydrochloric acid
0.1M sodium hydroxide solution
0.001M iodine solution, 40 cm^3
2 cm^3 dropping tube
1 cm^3 dropping tube
Three other dropping tubes (if available)
Two 250 cm^3 beakers, as water baths
Thermometer, 0–110 °C
Stopclock

Experiment 18.5c
The digestion of milk
Each student or pair of students will need:

Fresh milk, 50 cm^3
Commercial rennin solution (from a grocer), 3 cm^3
0.1M sodium oxalate solution, 3 cm^3
1 per cent pancreatin solution (as source of lipase), 2 cm^3
250 cm^3 beaker, as a water bath
Thermometer, 0–110 °C
Filtration apparatus

Procedure

Details for carrying out these experiments are given in the *Students' Book*.

A thermostatically controlled water bath set at 40 °C is useful.

To prepare starch solution make the dry starch into a thin cream with cold water and pour into *boiling* water. Boil briefly then allow to cool. Add no preservative for these enzyme experiments.

Urease will be found to cause the hydrolysis of urea only.

Amylase will function well in ordinary conditions but not in 0.1M acid or alkali. At 95 °C urease is denatured and will not function, while at 40 °C the rate of hydrolysis of starch is fast.

Rennin readily clots milk but fails to function when preheated to boiling or if calcium ions are absent. Lipase causes the hydrolysis of the fats in milk so that the solutions become acidic, but is denatured by heat and then fails to function.

Suggestions for homework

Reading the *Scientific American* offprint listed below.

Supporting material

Phillips, D. C. 'The three-dimensional structure of an enzyme molecule' *Scientific American* offprint No. 1055.

Summary

As a result of their work in this section students should know that enzymes are protein catalysts, and should have an appreciation of the factors that influence enzyme function. A detailed knowledge of lysozyme should not be required.

18.6 An examination of some plastics

Objectives

1   To examine qualitatively the mechanical and chemical properties of plastics.

2   To develop an awareness that the commercial usefulness of plastics has to take into account cost and processing as well as mechanical and chemical properties.

Timing

About four periods.

Suggested treatment

Students will obviously already have some knowledge of plastics. A good treatment would be to establish what is already known and then develop the discussion in terms of the interests expressed. Since different groups of students are likely to have different interests teachers will want to interweave the teaching of 18.6, 18.7, and 18.8 to suit their particular circumstances. For example, the stretching of plastic film which is carried out as experiment 18.6 (part 5) is considered again in relationship to molecular structure in 18.8.

Students are asked first to compare the relative merits of plastics and alternative materials for any household objects that occur to them.

Suitable examples might be combs (bone and nylon), electrical insulation (lead/natural rubber and PVC), washing-up bowls (enamelled iron and polythene or polypropylene). Well-worn examples of each type would form a useful display. In some cases the plastic will be clearly superior while in others quality has been lowered in favour of economy. This last point could form the basis of a useful discussion, or piece of written work.

The experimental work is a general examination of some plastics with the objective of choosing a plastic suitable for a carrier bag or a reagent bottle. Students might be asked to obtain plastics for testing but it will be difficult to assemble a proper range of materials of known composition. Books on the identification of plastics (see 'Supporting material') can be consulted and further suggestions are given in an appendix to this topic, which includes a listing of trade marks.

Chemical tests on plastics often disappoint students although the negative nature of the results in fact illustrates the advantageous chemical stability of these materials.

The mechanical tests may have already been considered as part of the students' physics course in which case the stretching tests will be a revision exercise. The strips of plastic must have clean cut edges (use a safety razor blade or a Stanley knife) otherwise tearing will occur under stress.

When polythene is pulled slowly and steadily a variety of effects can be observed. Initially the polythene corrugates along its length and then its width decreases in one or two places and the corrugations disappear or become very much smaller in these places. The force required to extend the material decreases when the thinning starts and continues at this smaller value as the reduction in width spreads along the specimen. It is particularly noticeable that when the polythene reaches the smaller width throughout its length, the force required to stretch it further and to break it becomes much higher. It is worth noting how elastic the material is at this stage.

The changes in temperature which occur at various stages in the pulling process can be felt by touching the polythene to the lips.

The above experiment can be repeated with a fresh piece of polythene but stop pulling when the specimen has gone 'hard' and before it breaks. Now pull the polythene across its width and see how easily it splits.

Pull another piece of polythene to the 'hard' stage and while it is under tension puncture it with a finger nail to see a split travel along the material. It is possible to shred the polythene into strands by doing this or by pulling it sideways while under tension.

When plastic strips subject to stress are heated they will increase in length but if natural rubber is observed closely it will be seen to contract. The elasticity of natural rubber can be described as an entropy effect and stretching rubber involves a decrease in entropy.

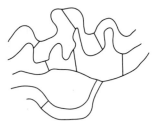

An unlikely conformation
(i.e. only obtained by stretching)

A probable conformation
(i.e. unstretched natural form)

Figure 18.6
The elasticity of natural rubber.

When a substance is heated its entropy increases and the molecular behaviour will change accordingly: in an atomic lattice increased atomic vibration; in rubber a contraction to a more probable conformation.

When plastics are heated there is a decrease in their intermolecular forces and the structure can therefore flow when stressed.

The fabrication of finished goods from plastic raw materials is described in the Nuffield Chemistry Background Book *Plastics* and also in the Nuffield film loop 'The manufacture of plastic articles'.

For a carrier bag, although other films would be stronger, polythene of 200 gauge (0.002 in. thick) is of adequate strength and is preferred to any other material on the score of cost (polythene bag – 1p, paper bag – 2p). An important point is that film for a carrier bag should be tested against a jerk-weight (of 10 kg) rather than a steady stress. The effect with polythene can be demonstrated very effectively by holding one end of a strip in a rigid support, attaching a large mass to the other end, and then allowing the mass to drop. The polythene will fracture with very little deformation, that is, it is nearly brittle.

The reason behind the choice of plastic bottles is mainly one of weight consideration, but other advantages such as resistance to breakage and flexibility of use present themselves in a favourable light. When considering the choice of polythene, one can normally divide chemicals into two groups.
    a   those which are incompatible with plastics, for example solvents, and
    b   those which are compatible.
Group (b) comprises the majority of dry chemicals, but there will always be a small group of dry materials which are not compatible with plastic. Polythene is an obvious choice as it is a 'pure' plastic, being formed by the polymerization of ethylene, whereas other plastics like PVC contain dibutyl phthalate plasticizers, which would cause contamination of many chemicals.

Both bottles and caps are produced from high density polythene which provides a fair degree of strength. Caps are injection moulded and bottles are formed by both injection and blow moulding in a split mould.

An additional example is the choice of packaging material for potato crisps. Some crisps, for example, are packed in orientated polypropylene film which has been coated with polyvinylidene dichloride. A competitive material is cellulose film but the plastic film has better resistance to puncture and is a good moisture barrier. The PVDC coating increases the odour barrier of the film and enhances the ability of the film to be run through machinery at high speed. Furthermore with a PVDC coating the bag can be heat sealed at 130 °C whereas orientated polypropylene film alone has to be heated to 160 °C to be heat sealed, at which temperature it shrinks and wrinkles.

The testing of plastics for particular applications is described in the Shell film 'The polyolefins' and in the Nuffield film loop 'Testing of plastics'.

Experiment 18.6
Choosing a plastic

Each student or pair of students will need:

Test-tubes and rack
Metal spatula
Potassium permanganate solution
Tetrachloromethane
Toluene
Plastics in the form of granules
Scouring powder (pumice)
Plastics in the form of approximately 100 gauge film (0.001 in. thick) as strips 25×1 cm
    500 g mass

access to:

A garden spade
A plot of ground

Procedure

Full details are given in the *Students' Book* and some of the more important results are described above. The action of heat on some typical materials is given in table 18.6a.

Plastic	Burning action	Flame colour	Odour	Special features
casein	heating necessary	yellow	burnt milk	—
nylon	heating necessary	blue, tipped yellow	burning vegetation	fibres can be drawn from melt
PVC	heating necessary	yellow, green at base	acrid	acidic fumes
polyethylene terephthalate	self-supporting	smoky, yellow	sweet	fibres can be drawn from melt
polypropylene	self-supporting	yellow, blue at base	as candle wax	forms clear liquid
polythene	self-supporting	yellow, blue at base	as candle wax	forms clear liquid
urea-formaldehyde	heating necessary	pale yellow, edged blue-green	fish-like	alkaline fumes

Table 18.6a
Action of heat on some plastics

Supporting homework

Writing an account of the relative merits of plastics and other materials for household objects.
Answering the questions in the Background Book *Plastics*.

Supporting material

Nuffield Chemistry Background Book *Plastics*.

Saunders, K. J. (1966) *The identification of plastics and rubbers*, Science Paper-backs, Chapman and Hall.

Nuffield film loops, 'The manufacture of plastic articles' and 'Testing of plastic film'.

16 mm film: 'The polyolefins', Petroleum Film Bureau, 4 Brook St, London W1.

Summary

As a result of their work in this section students should be aware of the chemical stability of plastics and also of the mechanical limitations (deformation, abrasion, temperature effects) to their use. They should be aware of cost as an important factor in evaluating the usefulness of a plastic and know something of the fabrication of objects from plastics.

In the next section students will be looking at some of the reactions which are used to make plastics.

18.7 The laboratory preparation of some plastics

Objective

To give students direct experience of some polymerization reactions and gain practice in the correct use of technical terms including monomer, polymer, addition or condensation polymerization, and thermoplastic or thermosetting resin.

Timing

Two periods.

Suggested treatment

Before attempting any of these experiments students should have discussed Carothers' views on polymerization given in section 18.8 and be familiar with the technical terms mentioned above.

Students cannot be expected to perform more than one or two experiments in the time available, but the results can be considered in a general class discussion. Alternatively some work could be done as a teacher demonstration.

A general warning should be issued about the pungent odour and caustic nature of many of the monomers. Laboratories with inadequate ventilation are advised to restrict the choice of experiments.

Experiment 18.7
The preparation of plastics

1   The polymerization of styrene
Each student or pair of students will need:

> Styrene
> 1,2-dibromoethane
> Lauroyl peroxide (or anhydrous aluminium chloride)
> Cotton wool
> Test-tubes 150×25 mm
> Measuring cylinder, 10 cm^3
> A water bath

2   Acrylamide polymerization
Each student or pair of students will need:

> Acrylamide
> Potassium or ammonium persulphate
> Beaker, 250 cm^3
> Thermometer, 0–110 °C
> 250 cm^3 'throw-away' container
> Protective polythene gloves

3   Formaldehyde resins
Each student or pair of students will need:

> a 40 per cent formaldehyde solution (Formalin)
>   Phenol
> 100 cm^3 'throw-away' container
> b 40 per cent formaldehyde solution (Formalin)
>   Urea
> 100 cm^3 'throw-away' container
> c This is an optional extra; see *Students' Book* for details.

4   Nylon rope trick
Each student or pair of students will need:

> Sebacoyl chloride or adipyl chloride
> 1,6-diaminohexane
> Tetrachloromethane
> 250 cm^3 beaker
> Crucible tongs
> The roller system illustrated in the *Students' Book* is optional

5   Casein
Each student or pair of students will need:

> 100 cm^3 milk (a pint bottle will provide 500 cm^3 of separated milk)
> 40 per cent formaldehyde solution (Formalin)
> Beaker, 250 cm^3
> Thermometer, 0–110 °C

Procedure
Full details are given in the *Students' Book*.

A supply of tin cans should be available for use as throw-away containers for these polymerizations, as laboratory glassware may prove impossible to clean. For an experiment such as 3(a) a shaped container like a rubber mould would add interest.

In experiment 3(c) the vacuum distillation can be omitted without much detriment to the product.

In experiment 4 nylon '610' is obtained from sebacoyl chloride and nylon '66' from adipyl chloride.

Supporting material
Shell International Petroleum Company Limited, booklet for teachers *Experiments in polymer chemistry*

Summary
At the end of this section students should have an understanding of the terms monomer, polymer, addition and condensation polymerization, thermoplastic and thermosetting resin. Their practical work should have given experience in the correct use of these terms.

They should also have enjoyed themselves.

18.8   The invention and discovery of synthetic polymers

Objectives
1   To develop an understanding of addition and condensation polymerization reactions, and the relationship of monomers to their corresponding polymers.
2   To develop an elementary understanding of the relationship between structure and properties in thermoplastics.
3   To describe the structure of stereoregular polymers.

Timing
About four periods.

Suggested treatment
For this treatment OP transparencies numbers 65 to 68 will be useful.

A knowledge of the material in this section is necessary for a proper understanding of the practical work of the two previous sections and as far as possible the teaching of the three sections should be interwoven.

It would be useful if students revised their work on X-ray diffraction and van der Waals' bonding before starting on this section.

Parts of the *Students' Book*, the introduction, the extract from Carothers' patent, and the account of the discovery of polythene, could be set as private reading. This reading matter is included because a subsidiary objective in this topic is to suggest that scientific awareness is not a simple attitude for it involves not only systematic thinking about problems but also the ability to profit from happy accidents. The stories of nylon and polythene stand in contrast while Ziegler's work involves a combination of logic and chance. It might also be suggested that only prepared minds make accidental discoveries – Dr Plunkett might well have discarded the gas cylinder whose contents had polymerized to PTFE.

The account of Carothers' work describes addition and condensation polymerization and asks what compounds might be suitable as monomers. Additionally the teacher could discuss here the formation of rigid thermosetting resins when polymer chains have functional groups available for cross-linking. Students should be familiar with this work before attempting the practical work of section 18.7. The Nuffield Chemistry film loop 2–10 'Plastics' may be helpful.

Carothers' work is used also to introduce the idea that polymer properties depend in an important way on molecular shape. The two reactions described can be demonstrated quite readily. One-tenth mole quantities (ethane-diol 5.6 cm^3, succinic acid 11.8 g, phthalic anhydride 14.8 g) can be melted together in boiling tubes then heated for an hour in an oil bath at 180 °C. When left overnight the products should be nearly as described in the *Students' Book*. Students should build ball-and-spoke models of the monomers (separate groups of students for each monomer avoids confusion) and then combine them in the polymers. Attention might be drawn to the elimination of the components of water which occurs. It will be found that the succinate ester can form a neat linear polymer but in the phthalate ester the benzene rings may stick out from the polymer chain (figure 18.8a).

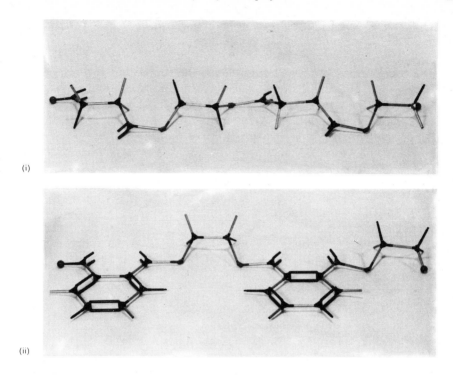

Figure 18.8a
Models of polymers.
i Model of the ethylene succinate polymer.
ii Model of the ethylene phthalate polymer.

It should then be quite easy to see that if a carboxyl group is moved from its *ortho* position to the *para* position on the benzene ring a neat linear polymer will result. Students will probably not know that this is polyethylene terephthalate (Terylene fibre and Melinex film). The strength associated with this neat linear polymer is best appreciated by attempting to stretch Melinex film.

It is considered that model building is a most important activity if students are to grasp the significance of molecular shape.

With the information from model building it should be apparent that polymers of neat molecular shape are crystalline and can possess considerable mechanical strength. Such polymers are likely to be useful commercially.

The idea that many plastics are crystalline (moulded nylon about 40 per cent, low density polythene about 60 per cent) will surprise and puzzle many students.

The evidence of the X-ray diffraction photographs (*Students' Book* figures 18.8a, b, c, and f) should be sufficient to establish the point.

The cold drawing of plastics to orientate their crystalline regions greatly increases their mechanical strength. The X-ray diffraction photographs of nylon in the *Students' Book* figure 18.8a indicate the changes that occur. The X-ray fibre photograph of undrawn nylon resembles a powder photograph, indicating the crystalline but non-orientated nature of the material. But when drawn a good fibre photograph is obtained indicating that the crystalline regions have now been orientated.

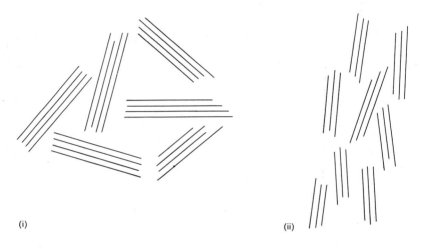

(i)   (ii)

Figure 18.8b
Orientation of crystalline regions in a polymer by cold drawing.
i Undrawn
ii Drawn

Nylon tubing is partially orientated by the manufacturing process (extrusion) and should be classed as anisotropic. During the cold drawing process magnificent colouring effects can be observed if the tubing is held at 45 °C between crossed polaroids.

The effects between crossed polaroids are even more marked when polythene is cold drawn. It will be found that polythene strips cut in the 'machine direction' can be cold drawn, but at right-angles the strips will stretch and break. This is due to partial orientation in the manufacturing process.

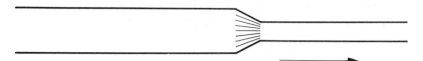

Figure 18.8c
The necking down effect in cold drawing.

The coloured effects seen when stretching film between crossed polaroids are due to changes in structure which result in variation in the refractive index of the material. No further interpretation is intended; the effects are worth observing for their intrinsic attractiveness.

This work on the cold drawing of nylon and polythene is an extension of the practical work in section 18.6.

The work of Ziegler and Natta is not easy to grasp. Emphasis is best placed on the greater crystallinity, and hence mechanical strength, likely to result from stereoregular molecular shapes. A model of an isotactic structure (figure 18.8d) is worth building with a space-filling model set. Each carbon atom is rotated 120° relative to the previous atom so that eventually there are three separate straight lines of hydrogen atoms along the carbon chain. The pattern is seen clearly if the side chain methyl groups are left off.

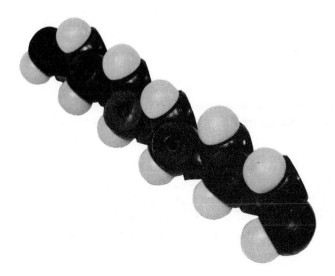

Figure 18.8d
A space-filling model of an isotactic structure (without side chains).

A proposed mechanism for Ziegler catalysts is presented in figure 18.8e and could be used to answer students' questions if they arise, or they could be referred to the chapter on catalysts in *The chemist in action*.

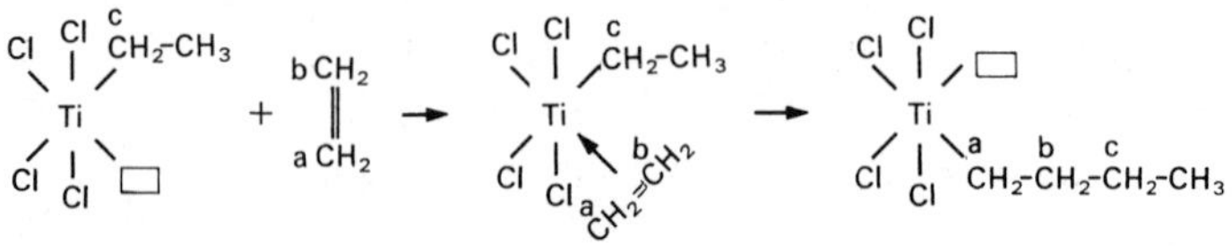

Figure 18.8e
The polymerization of olefines using titanium tetrachloride – one of the Ziegler-Natta catalysts. First the vacant space absorbs the molecule which subsequently undergoes rearrangement to form new carbon-carbon bonds.

Finally, attention is drawn to the dramatic difference in mechanical properties between the stereoregular materials, natural rubber and gutta percha which differ structurally only in being *cis* and *trans* respectively across the double bond in the molecular chain (*Students' Book*, figure 18.8g). If at all possible the properties of gutta percha should be demonstrated.

Demonstration materials
The teacher will need:

Ethane-1,2-diol
Succinic acid
Phthalic anhydride
Test-tubes, 150×25 mm
Thermometer, 0–250 °C
Oil bath
Space-filling model set
Strips of natural rubber and gutta percha

Each student or pair of students will need:

Ball-and-spoke model set
Strips of Melinex, nylon and polythene film
Sharp knife to cut strips of film
Polaroid assembly (Topic 8)

Supporting homework
Revision of X-ray diffraction (Topic 8).
Revision of van der Waals' bonding (Topic 10/11).
Reading the chapter on Polymers in *The chemist in action*.

Supporting material
Nuffield Chemistry Film Loop 2–10 'Plastics'.

Summary
As a result of their work in this section students should know about polymerization reactions and appreciate that there is a relationship between mechanical properties, crystallinity, and molecular shape in plastics.

Appendix I
Obtaining samples of plastics

The following materials are required:

  Low density (LD) polythene granules (old tubing could be cut up)
  Nylon granules (Griffin and George Limited)
  PVC granules (cut up tubing)
  Polypropylene granules (household goods branded Propathene or plastic boxes with integral moulded hinge, like a film loop box)
  Casein granules (buttons, smell of burnt milk when heated)
  LD polythene film (the common material, food bags from chain stores)
  Nylon film (Portland Plastics Limited)
  Polyethylene terephthalate film (recording tape or Melinex from J. M. Wiley Limited)
  PTFE film (F. R. Warren and Company Limited)
  Natural rubber strip (a model shop)
  Gutta percha film (used for binding cut flowers, from horticultural and florist sundriesmen)

The yellow pages of the telephone directory will indicate local suppliers of plastics, where small offcuts of other materials might be bought. The specialist firms listed above are as follows:

Portex Ltd,	Autoclavable nylon film, $\frac{1}{2}$ inch width
Hythe,	tubing, B gauge (0.001 inch)
Kent	Price: about 15p per hundred feet
J. M. Wiley Ltd,	Melinex film, Type 0, 50 gauge (0.0005 inch),
Victoria Road,	metallized
South Ruislip,	Price: about £3
Middlesex.	
F. R. Warren and Co. Ltd,	PTFE skived tape, 200 gauge (0.002 inch),
79 Ashley Down Road,	5×1 ft
Bristol 7.	Price: about 50p

Appendix II
Trademarks for plastics

Abstrene	Acrylonitrile-butadiene-styrene (BXL Ltd)
Acrilan	Acrylic fibre (Chemstrand)
Akulon	Nylon 6 and 66 (AKU, Holland)
Alathon	Polythene (Du Pont)
Alkathene	LD Polythene (ICI)
Argosy	Melamine-formaldehyde tableware (J. S. Peress Ltd)
Astralit	PVC sheet (Dynamic-Nobel AG, Germany)
Bakelite	Plastics raw materials, including phenol-formaldehyde (BXL Ltd)
Beetle	Urea- and melamine-formaldehyde moulding materials (BIP Chemicals Ltd)
Bexan San	Styrene-acrylonitrile (BXL Ltd)
Bextrene	Polystyrene (BXL Ltd)
Blue C nylon	Nylon 66 fibre (Chemstrand)
Breon	PVC and nitrile rubber (BP Chemicals (UK) Ltd)
Bri-nylon	Nylon 66 fibre (ICI Fibres Ltd)
Butakon	Butadiene and nitrile rubbers (ICI)
Cariflex	Synthetic rubbers (Shell)
Carina	PVC (Shell)
Carinex	Polystyrene (Shell)
Carlona	Polyolefins (Shell)
Cellophane	Cellulose film (British Cellophane Ltd)
Celon	Nylon 6 fibre (Courtaulds)
Cirrus	Expanded PVC-coated fabric (Bernard Wardle Ltd)
Cobex	Rigid PVC sheet (BXL Ltd)
Contact	Self-adhesive PVC sheet (Storey Bros. & Co. Ltd)
Corvic	PVC polymer (ICI)
Courtelle	Acrylic fibre (Courtaulds)
Crimplene	Polyethylene terephthalate bulked fibre (ICI Fibres Ltd)
Dacron	Polyethylene terephthalate fibre (Du Pont)
Darvic	Rigid PVC sheet (ICI)
Diakon	Acrylic moulding powder (ICI)
Dralon	Acrylic fibre (Farbenfabriken Bayer AG, Germany)
Enkalon	Nylon 6 fibre (British Enkalon Ltd)
Epikote	Epoxy resins (Shell)
Erinoid	Casein (BP Chemicals (UK) Ltd)
Fablon	Self-adhesive PVC film (Commercial Plastics Ltd)
Fertene	Polythene (Montecatini Edison, Italy)

Fiesta	Melamine-formaldehyde tableware (Brooes and Adams Ltd)
Flovic	Vinyl copolymer sheet (ICI)
Fluon	PTFE (ICI)
Formica	Melamine-formaldehyde laminates (Formica Ltd)
Gaydonware	Melamine-formaldehyde tableware (BIP Chemicals Ltd)
Geon	PVC (B. F. Goodrich Chemical Co. USA)
Kayfoam	Flexible polyurethane foams (Kay Bros. Plastics Ltd)
Kematal	Acetal copolymer (ICI)
Kotina	Expanded polystyrene wall covering (Poron Insulation Ltd)
Lexan	Polycarbonates (General Electric Co. USA)
Lucite	Acrylic sheet and moulding powder (Du Pont)
Lustran 1	Styrene-acrylonitrile (Monsanto)
Lustran A	Acrylonitrile-butadiene-styrene (Monsanto)
Lustrex	Polystyrene (Monsanto)
Makroware	Polycarbonate housewares (Industrial Mouldings (Warwick) Ltd)
Maranyl	Nylons (ICI)
Marlex	Polyolefins (Phillips Petroleum, USA)
Megaprufe	Expanded polystyrene wall covering (Omega Plastics Co. Ltd)
Melaware	Melamine-formaldehyde tableware (Ranton & Co. Ltd)
Melinex	Polyethylene terephthalate film (ICI)
Mellowear	PVC-coated fabric leathercloth (Mellowhide Products Ltd)
Melmex	Melamine-formaldehyde (BIP Chemicals Ltd)
Montopore	Expanded polystyrene (Monsanto)
Montothene	Ethylene-vinyl acetate (Monsanto)
Moplen	Polypropylene (Montecatini Edison, Italy)
Mylar	Polyethylene terephthalate film (Du Pont)
Oroglas	Acrylic sheet and moulding powder (Lennig Chemicals Ltd)
Perlon	Nylon 6 (Glanzstoff AG, Germany)
Permaware	Melamine-formaldehyde tableware (VIP Ltd)
Perspex	Acrylic sheet (ICI)
Plexiglas	Acrylic sheet and moulding powder (Rohm and Haas, Germany)
Polyflex	Polystyrene film (Monsanto)
Propafilm	Polypropylene film (ICI)
Propathene	Polypropylene (ICI)

Qiana	Polyamide (Du Pont)
Rexine	Nitrocellulose-coated fabrics (ICI)
Rigidex	High-density polythene (BP Chemicals (UK) Ltd)
Rilsan	Nylon 11 and 12 (Aquitaine-Fisons Ltd)
Rockite	Phenolic moulding powder (BP Chemicals (UK) Ltd)
Royalite	Acrylonitrile-butadiene-styrene i.e. ABS sheet (North British Rubber Co. Ltd)
Saran	Polyvinylidene chloride (Dow Chemical Co.)
Styrocell	Expanded polystyrene (Shell)
Styron	Polystyrene (Dow/Distrene Ltd)
TPX	Poly-4-methylpent-1-ene (ICI)
Teflon	PTFE (Du Pont)
Terylene	Polyethylene terephthalate fibre (ICI)
Transpex	Extruded acrylic sheet (ICI)
Tricel	Cellulose triacetate fibre (Courtaulds)
Tyril	Styrene-acrylonitrile (Dow/Distrene Ltd)
Ultramid	Nylon (BASF, Germany)
Ulstron	Polypropylene filament yarn (ICI Fibres Ltd)
Velbex	Flexible PVC sheeting (BXL Ltd)
Viclan	Polyvinylidene chloride (ICI)
Vincel	Polynosic fibre – a modified rayon (Courtaulds)
Visqueen	LD Polythene sheeting and film (British Visqueen Ltd)
Vulkide A	Acrylonitrile-butadiene-styrene sheet (ICI)
Vulkide B	Polypropylene sheet (ICI)
Vynide	PVC coated fabrics (ICI)
Warerite	Melamine-formaldehyde laminates (BXL Ltd)
Welvic	PVC compounds (ICI)
Xylonite	Celluloid sheet (BXL Ltd)
Zytel	Nylon (Du Pont)

Topic 19
Some p-block elements

Objectives

To study some aspects of the chemistry of nitrogen and sulphur, using ideas learned from redox potentials, equilibria, kinetics and other topics earlier in the course. Only a small portion of the chemistry of these elements can be covered in two weeks; teachers may wish to deal with aspects of chemistry of these elements other than those mentioned here. But it is also a useful opportunity to revise information about these elements which has already been encountered.

Timing

Two weeks.

Content

19.1   Nitrogen
19.2   Sulphur

Each section starts with an introduction to the chemistry of the element and this is followed by structured practical work.

It is most important to note that the practical work is structured. Students should answer the questions, *as they carry out the practical work*. Help and discussion will be needed from the teacher on a personal basis, in small groups, or in class discussion.

19.1 **Nitrogen**

Objective

To study some of the chemistry of nitrogen and its compounds in which the element has either a positive oxidation number, or an oxidation number of -3, using ideas already encountered in the course.

Timing

Six to seven periods.

Suggested treatment

For this treatment OP transparencies numbers 79, 114, and 117 will be useful.

Students can be introduced to the chemistry of nitrogen by discussion about those compounds of nitrogen which they have already met – ammonia, nitrogen dioxide, nitrites, and nitrates amongst others. The oxidation numbers of nitrogen in these compounds should be calculated and then the students can be referred to the table of all the oxidation numbers of nitrogen in the *Students' Book*.

The further usefulness of oxidation numbers should now be discussed along the lines of the material in the *Students' Book*. This material is intended as notes for the students' reference after the discussion.

The reason for five electrons being available for bonding should be discussed, and also the fact that nitrogen normally achieves an octet of electrons in compounds (and in molecular nitrogen). But it is useful to point out that this does not always occur: both in nitrogen monoxide, NO, and in nitrogen dioxide, NO_2 (but not in dinitrogen tetroxide) the number of electrons in the second shell is seven. It is also useful to point out that the promotions of electrons which are discussed do not actually occur before reaction, just as the various stages in a Born-Haber cycle do not occur in turn. But such ideas are the basis of a useful model. During this discussion ideas concerning electronic structure and ionization energies, oxidation numbers, and covalent and dative covalent bonds should be revised.

The structure of the nitrogen molecule and its stability should be discussed. This can lead to a consideration of the special methods required to 'fix' atmospheric nitrogen in a form in which it can be assimilated by plants.

Experiment 19.1
An experimental investigation of the chemistry of nitrogen

Each student will need:
Test-tubes
Dropping pipettes

access to:
Aluminium foil or turnings
Ammonium chloride
Copper turnings
Copper(II) nitrate
Devarda's alloy
Ice
Nitric acid, concentrated
Potassium chloride
Potassium or sodium nitrate
Potassium or sodium nitrite
Sulphuric acid, concentrated

and to solutions of:
0.2M sodium hydroxide
0.2M ammonia
0.2M solutions containing the following ions
    lead
    calcium
    zinc
    copper(II)
    magnesium
bench potassium iodide
bench potassium permanganate
bench dilute hydrochloric acid
bench sodium hydroxide
saturated solution of H_2S in water (freshly prepared)

Procedure
The practical work starts off with the question, 'What nitrogen compounds can be made from nitric acid?'

Students should be reminded of how to use the redox potential charts to find whether a reaction is possible, and that this only indicates whether a reaction is *possible* – it may occur at an immeasurably slow rate.

It is important to stress that if instructions are given exactly, they should be followed exactly: *a few drops* of concentrated nitric acid and *three or four* copper turnings prevents too much nitrogen dioxide being produced.

In the first section, (a) and (b), concentrated nitric acid and copper give dinitrogen tetroxide and it might be expected from the equations that more dilute nitric acid would give nitrogen monoxide and/or dinitrogen monoxide and that even more dilute acid would give nitrogen and/or ammonium ions. Before they carry out (b), students should be shown a gas jar of nitrogen monoxide, and what happens when it is mixed with a gas jar of air. What might happen when NO is produced at the bottom of a test-tube full of air should then be discussed.

With 7M acid, a very light brown mixture of gases is produced which would indicate that NO_2, NO, and possibly N_2O have been formed.

In the next section, (c) and (d), ammonia and hydrogen are obtained from a nitrate in alkaline solution, using Devarda's alloy. The $E^{\ominus}$ values indicate that neither copper nor aluminium will bring about this reaction under standard conditions. But as the $E^{\ominus}$ values are close (-2.87 V and -2.35 V) for reduction by aluminium, it is more likely that, under suitable conditions, aluminium rather than copper will reduce nitrate to ammonia. Although aluminate is more correctly $H_2AlO_3^-$ (aq), the simpler formula is used in the equation

$$5OH^-(aq) + 2H_2O(l) + 3NO_3^-(aq) + 8Al(s) \rightarrow 3NH_3(aq) + 8AlO_2^-(aq)$$

It can be seen that the reaction is very sensitive to hydroxide ion concentration and that increase of this should cause the reaction to occur.

Practical investigation confirms these predictions.

The equation for the production of hydrogen is:

$$2OH^-(aq) + 2H_2O(l) + 2Al(s) \rightarrow 2AlO_2^-(aq) + 3H_2(g)$$

In the next section, (e) and (f), H_2S is oxidized by nitric acid, first to sulphur and then to sulphate. Iodide ions are also oxidized by nitric acid, iodine being formed.

Nitrous acid is then studied in (g) and (h). In acid solution disproportionation occurs to NO_3^-(aq) and NO_2(g). $E^{\ominus}$ values indicate that this reaction cannot occur in alkaline solution. As nitrogen cannot have an oxidation number higher than that in nitrate, similar disproportionation cannot occur with nitric acid.

From $E^{\ominus}$ values nitrite ions should react both with permanganate and iodide ions, in the one case being oxidized to nitrate, and in the other being reduced to nitrogen monoxide or dinitrogen monoxide.

The next section, (j), is concerned with the decomposition of nitrates: sodium or potassium nitrate to the nitrite which gives NO_2 with acid, and copper(II) nitrate to the oxide.

The final section, (k), (l), (m), (n), and (p) deals with nitrogen with an oxidation number of -3. Ammonium chloride solution is acidic because of the equilibrium

$$NH_4^+(aq) \rightleftharpoons NH_3(aq) + H^+(aq)$$

Thus addition of magnesium gives hydrogen. No such equilibrium occurs in potassium chloride solution which is therefore neutral.

The pH of 0.2M NaOH solution is about 13, that of 0.2M NH_3 solution about 9, indicating that the respective [OH$^-$(aq)] are about 0.2M and about 0.00001M. Sodium hydroxide exists as ions so 0.2M NaOH is 0.2M in OH$^-$(aq). Hydroxide ions are formed when ammonia is dissolved in water according to the equilibrium

$$NH_3(aq) + H_2O(l) \rightleftharpoons NH_4^+(aq) + OH^-(aq)$$

Sodium hydroxide gives precipitates with solutions of Pb^{2+}(aq), Ca^{2+}(aq), Zn^{2+}(aq), and Cu^{2+}(aq), lead and zinc hydroxides dissolving if sufficient sodium hydroxide solution is added, to give anions. Ammonia solution does not give a precipitate with calcium ion solution as the [OH$^-$(aq)] is not high enough for the solubility product of calcium hydroxide to be reached, but it gives a precipitate with the other three solutions, copper(II) hydroxide and zinc hydroxide dissolving to give complex ions.

When ammonium chloride is added to a solution of ammonia, the [OH$^-$(aq)] is decreased and [NH_3(aq)] increased as the equilibrium:

$$NH_3(aq) + H_2O(l) \rightleftharpoons NH_4^+(aq) + OH^-(aq)$$

is shifted to the left. Thus this solution does not precipitate magnesium hydroxide when added to a solution of magnesium ions (whereas ammonia solution does) because the hydroxide ion concentration is lower and more NH_3(aq) is available for complex formation.

Finally the teacher should discuss why nitrogen has to be 'fixed' and how this can be done. The Haber process should be discussed briefly and how the ammonia so produced is converted into nitric acid can be demonstrated in the laboratory. A brief study of this industrial process can be illustrated by some short demonstrations.

The highly exothermic nature of the catalytic oxidation process can be demonstrated using the apparatus in figure 19.1a.

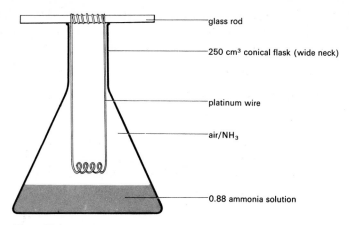

Figure 19.1a

The platinum wire is heated to red heat and inserted in the conical flask. It continues to glow spontaneously. 24 gauge copper wire can be used instead of platinum, but in this case the heat evolved in the reaction melts the copper which forms the deep blue $Cu(NH_3)_4^{2+}$ ion in the ammonia solution. Heavier gauge copper wire fails to melt and glows as for platinum.

The production of brown NO_2 gas as an intermediate in the industrial process can be illustrated using the apparatus shown in figure 19.1b. It is sensible to use a safety screen with the apparatus.

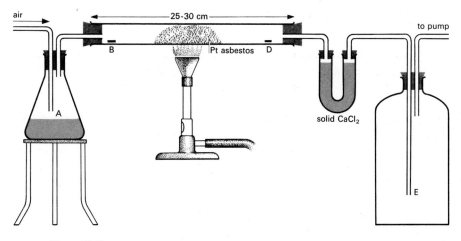

Figure 19.1b

The platinized asbestos must be dried overnight in an oven. At either end of the tube a 5 cm length of blue litmus paper is placed. The flask A contains 6M ammonia solution. The air inlet tube must not dip into the ammonia solution. The receiver E may be a flask or aspirator bottle, but it should be large and should be set against a white background so that the brown fumes are visible. The combustion tube is warmed, and when the platinum is red hot the water suction pump is turned on. Soon brown fumes appear in E and the litmus D turns red. The rate of suction of air may require regulation. The tube of calcium chloride absorbs excess ammonia.

In the industrial process, solution of the NO_2 in water in the presence of atmospheric oxygen gives nitric acid:

$$4NO_2(g) + O_2(g) + 2H_2O(l) \rightarrow 4HNO_3(aq)$$

Summary

Students should be more able to use ideas encountered earlier in the course, and should know something more about the chemistry of nitrogen.

19.2 Sulphur

Objectives

To study some of the chemistry of sulphur, and of its compounds in which a variety of oxidation numbers of the element are seen, using ideas already encountered in the course.

Timing

Six to seven periods.

Suggested treatment

For this treatment OP transparencies numbers 44, 79, 114, and 118 will be useful.

Students can be introduced to the chemistry of sulphur by discussion about those compounds of sulphur which they have already met – sulphates, sulphites, sulphur dioxide, and hydrogen sulphide amongst others. The oxidation numbers of sulphur in these compounds should be calculated and the students can be referred to the table of all the oxidation numbers of sulphur given in the *Students' Book*.

The reason for six electrons being available for bonding should be discussed. It should be pointed out that sulphur has an octet of electrons in H_2S but in SO_3^{2-}, this is expanded to ten, and in SO_4^{2-} to twelve electrons. The oxidation numbers of ions containing S—S bonds should be discussed, and also the situation in persulphate. The material in the *Students' Book* is intended as notes for students' reference after the discussion.

Experiment 19.2a
Reactions of some sulphur compounds

Each student will need:

   Test-tubes
   Dropping pipettes

access to:

   Sodium sulphite (anhydrous) (preferably a new bottle)
   Sodium thiosulphate
   Sodium sulphide
   Finely ground *roll* sulphur
   Potassium peroxydisulphate (persulphate)

and to solutions of:

   bench dilute hydrochloric acid
   bench sodium hydroxide solution
   bench barium chloride solution

Procedure

The practical work starts with an investigation into the stability of various compounds of sulphur in which the element exhibits its principal oxidation numbers.

In sections (a) and (b) sulphite ions are in equilibrium with sulphur dioxide in acid solution, and on heating the equilibrium is disturbed and SO_2 is given off. Disproportionation is possible, but as hydrogen ions are involved on the right-hand side of the equation this reaction is prevented by the presence of acid.

$$4SO_2(aq) + 3H_2O(l) \rightarrow 2SO_4^{2-}(aq) + S_2O_3^{2-}(aq) + 6H^+(aq)$$

Thus when acid is added to a sulphite and heated, SO_2 is given off. Disproportionation is possible in alkaline solution, but after heating, acidifying and adding barium chloride solution, no precipitate is formed so no sulphate has been produced – the reactions are prevented kinetically.

Disproportionation cannot occur with sulphate ions as sulphur here has its highest (normal) oxidation number.

In section (c) disproportionation could occur with thiosulphate ions giving sulphur and either tetrathionate ions and/or sulphur dioxide and/or sulphate ions. On heating sodium thiosulphate with acid, sulphur is formed, and sulphur dioxide evolved. Disproportionation cannot occur in alkaline solution.

In sections (d) and (e) disproportionation cannot occur with sulphide ions as sulphur here has its lowest oxidation number. In acid solution, H_2S is formed on warming, due to the fact that H_2S is a gas and so, when given off, the equilibrium:

$$2H^+(aq) + S^{2-}(aq) \rightleftharpoons H_2S(aq)$$

is displaced to the right.

In section (f), disproportionation cannot occur with sulphur in acid solution but could give sulphide and sulphite ions and/or sulphate and/or thiosulphate ions in alkaline solution. After boiling sulphur with alkali, if the solution is acidified, sulphur is formed showing that thiosulphate has been formed.

The thermal stability of some ions and molecules are then investigated in (g). Persulphate, thiosulphate, and sulphur have S—S or O—O bonds which may break on heating and these substances decompose on heating (sulphur rings are broken when sulphur is heated).

Experiment 19.2b
The oxidation of sulphite ions in solution

Each student will need:

> Burette
> Pipette
> 250 cm^3 conical flask

access to solutions of:

> 0.05M iodine in 0.15M potassium iodide solution
> 0.1M sodium sulphite from a new bottle. (10 mg dm^{-3} of disodium salt of EDTA
> can be added to prevent aerial oxidation)
> 0.1M cerium(IV) sulphate
> The most convenient way to make up this solution is to dilute ampoules of
> concentrated volumetric solution appropriately.
> Alternatively, at rather less expense, but with more trouble, a technical cerium salt can
> be used. 66 g technical cerium(IV) sulphate (ceric sulphate) which contains about
> 50 per cent cerium(IV) sulphate will give 1 dm^3 of an approximately 0.1M solution.
> 66 g of the technical salt is dissolved in a mixture of 30 cm^3 concentrated sulphuric
> acid and 500 cm^3 water by stirring. If necessary, it is filtered and then diluted to 1 dm^3
> with water. The solution must then be standardized against Fe(II) solution.
> 0.02M potassium permanganate
> Bench dilute sulphuric acid

Procedure
Iodine solution, cerium(IV) sulphate solution, and potassium permanganate solution are separately titrated with sodium sulphite solution (in which the main species present is $SO_2(aq)$ at pH 0.5, at which the titrations are carried

out). It is found that iodine oxidizes S(IV) to S(VI) (oxidation number change = 2). The oxidation number change for cerium(IV) is *about* 1.4 and for permanganate *about* 1.8, indicating that both sulphate and dithionate have been formed.

A possible model to explain the production of dithionate assumes that one electron is transferred to the oxidant at a time, the SO_2 forming a HSO_3 radical:

$$SO_2(aq) + H_2O(l) \rightarrow HSO_3(aq) + H^+(aq) + e^-$$

If the next particle with which this HSO_3 collides is another HSO_3, then they may react to form dithionate:

$$2HSO_3(aq) \rightarrow 2H^+(aq) + S_2O_6^{2-}(aq)$$

or may disproportionate to give sulphur(IV) and (VI)

$$2HSO_3 \rightarrow SO_2(aq) + SO_4^{2-}(aq) + 2H^+(aq)$$

If the next particle is another oxidant species, then oxidation may occur to sulphate:

$$H_2O(l) + HSO_3(aq) \rightarrow SO_4^{2-}(aq) + 3H^+(aq) + e^-$$

If dithionate is stable to further oxidation, then such a mechanism would account for the production of dithionate. In this case, if oxidant is added to the sulphite solution, we would expect more dithionate formed as it is more likely that the HSO_3 should meet another HSO_3 under these conditions than if the sulphite is added to the oxidant. This may be tried.

No dithionate is formed in the case of oxidation by iodine. This could be due to another mechanism in which both electrons are transferred at once or it could be due to the fact that dithionate is easily oxidized by iodine. Whether this is so can be easily checked by adding 10 cm^3 of each oxidant solution to 10 cm^3 bench dilute sulphuric acid plus 25 cm^3 approximately 0.05M sodium dithionate separately and noting the time for the oxidant to be used up in each mixture.

Summary
Students should have had more practice in using ideas encountered earlier in the course, and should know something more of the chemistry of sulphur.

Appendix 1
Nomenclature, units, and abbreviations

Introduction
We live in times in which everything seems to be changing, and the names of chemicals, and the units of measurement, are no exceptions to this generalization. Changes in these areas make it possible to handle and understand the rapidly increasing amount of knowledge available in as convenient a manner as possible.

Although a little difficult at first for teachers, school trials have shown that the use of the internationally-agreed systems of nomenclature and units described here presents no difficulties to the students. Indeed, experience has shown that the intended advantages of clarity and ease of comprehension are gained by students using these more rational systems.

Nomenclature
The naming of chemicals in this course follows the recommendations of the International Union of Pure and Applied Chemistry (IUPAC). The main principles are set out in the Nuffield Chemistry *Handbook for Teachers*, chapter 2, pages 34 to 43, to which reference should be made. Only a brief outline will be given here.

The simpler inorganic compounds of the non-metals are named using prefixes corresponding to the numbers of each atom in the empirical formula. Thus

N_2O is dinitrogen monoxide (not nitrous oxide)
NO is nitrogen monoxide (not nitric oxide)
P_2O_3 is diphosphorus trioxide (not phosphorous oxide)

Compounds of metals are named following the Stock notation. In this, the oxidation number of the metal is given in the name as a Roman numeral, whenever that metal can have more than one oxidation number. Thus

$FeSO_4$ is iron(II) sulphate (not ferrous sulphate)
$Fe_2(SO_4)_3$ is iron(III) sulphate (not ferric sulphate)
PbO_2 is lead(IV) oxide (not lead dioxide)

The last example would be pronounced as 'lead four oxide'. Exceptions to this notation arise when a metal has two oxidation numbers in the same compound. In these cases it may be more convenient to use the system described above for non-metals. For example, Pb_3O_4 is named trilead tetroxide, rather than the more cumbersome dilead(II) lead(IV) oxide.

The rules for naming the simpler organic compounds systematically are set out in the Nuffield Chemistry *Handbook for Teachers* as previously mentioned, and are also explained in some detail in the *Students' Book* as an appendix to Topic 9. A programmed text *Names and formulae of carbon compounds* is also available.

In general, systematic names are adhered to throughout these texts, but trivial names may be used in two possible circumstances:

1   Where use of a systematic name would be unduly cumbersome (as, for example, in the case of methyl orange) or even impossible (as in the cases of, say, chlorophyll and insulin).

2   Where a common compound is widely known by a trivial name; for example, acetic acid is used rather than the systematic ethanoic acid.

In cases of doubt over nomenclature, the reader is referred to the *Handbook for Chemical Society Authors*, published by the Chemical Society (1961), or to *An introduction to chemical nomenclature* by R. S. Cahn, published by Butter-worths (1968).

Units

The internationally accepted system of units known as the *Système International d'Unités*, or SI units, is followed in all Nuffield Advanced Science publications. This is a coherent system based on the six units of metre (symbol m), kilo-gramme (kg), second (s), ampere (A), kelvin (K), and candela (cd), and a proposed seventh unit, the mole (mol).

The principal implications for chemists that arise in these texts are now summarized. For full details the reader is referred to *Physico-chemical quantities and units* by M. L. McGlashan (1968), Royal Institute of Chemistry Monographs for Teachers No. 15.

Multiples of units – Powers of ten are indicated by prefixes to the names of the units: for example, a thousand metres is known as a kilometre (km). Preference is given to intervals of 10^3; thus for *preference* a length should be quoted as 125 millimetres (mm) rather than 12.5 centimetres (cm). Centimetres are not, however, forbidden, and may be used if desired.

The nanometre – When recording measurements such as the wavelength of light, or interatomic distances, a widely used unit has been the Ångstrom, which is 10^{-10} m. The preferred SI unit is 10^{-9} m, known as a nanometre (nm). In this course, such distances are quoted in nanometres: 1 nm = 10 Å. It is thought desirable that students should be aware of the Ångstrom unit, as they will meet it in other books, and its use has not been entirely excluded from the *Students' Books*.

The cubic decimetre and cubic centimetre – The cubic metre (m^3) is an inconveniently large unit of volume for practical purposes, and chemists will normally use the cubic decimetre (dm^3) and the cubic centimetre (cm^3). The word litre is now taken as a special name for the cubic decimetre. Although this name may be used in everyday speech it is an unnecessary extra word, and neither it nor the 'millilitre' are used in this course.

The mole – The unit of amount of substance is the mole, symbol mol, and this is defined as the amount of substance which contains as many elementary units as there are atoms in 0.012 kg of carbon-12. A solution containing one mole dissolved in sufficient solvent to make one cubic decimetre of solution is called a molar solution, symbol M. Thus M NaCl contains 1 mol dm^{-3} or 58.5 g dm^{-3}.

The kelvin – The unit of thermodynamic temperature is the kelvin (*not* 'degree kelvin'), symbol K. For practical purposes the texts often refer to temperatures in degrees Celsius (centigrade), symbol °C. 0 °C = 273 K.

The joule – The SI unit of energy is the joule, symbol J and all heat measurements are therefore given in joules, and not calories. 1 calorie = 4.1868 J.

Pressure – As mentioned above, the unit of energy is the joule, J. Force is expressed in joules per metre, J m^{-1} or *newtons*, N. Pressure is expressed in newtons per square metre, N m^{-2}. For practical purposes it is sometimes convenient to talk about atmospheres, and to measure pressures in millimetres of mercury, mmHg, and all three units are used at various places in this book.

1 atmosphere = 760 mmHg = 101 325 N m^{-2}

Abbreviations

The principal abbreviations or symbols for units have already been mentioned; others are explained in the text. State symbols are used in many equations; these have the following meanings:

(s) = solid	(g) = gas
(l) = liquid	(aq) = dissolved in water

Appendix 2
Apparatus, materials, and reagents

Apparatus
In this *Teachers' Guide* the apparatus required to perform each experiment is listed before the experiment is described. Some experiments require apparatus that is not generally available, and for others, it is possible to construct apparatus as an alternative to that which is commercially available. The construction of this apparatus is described here.

1 Syringes and accessories
Gas syringes*

Exelo Glass 100 cm³ with 1 cm³ graduation marks.	W. G. Flaig & Co., Exelo Works, Margate Road, Broadstairs, Kent	About £4 each
Summit Glass 100 cm³ with 5 cm³ graduation marks. Centre-fitting or side-fitting 'luer' nozzle.	Smith and Nephew Group, Southall's Ltd, Bessemer Road, Welwyn Garden City, Herts.	About £2.70 each
Stylex Plastic 50 cm³ with printed 1 cm³ graduations.	Willen Bros. Ltd, 44 New Cavendish Street, London W1	About 30p each (cheaper in bulk)
2 cm³ glass. Many types available – should be two-piece and all glass with 'luer' fitting: S54–550	Griffin and George Ltd, Ealing Road, Alperton, Wembley, Middlesex. and Smith and Nephew Group	About 20p each at the dozen rate. Dearer if bought singly.

Hypodermic needles

'luer' fittings: S54–598	Griffin and George Ltd	About 30p per dozen, size 1 or 2.
'Star' type	Smith and Nephew Group	

Notes:

1 If 'Summit' 100 cm³ syringes are used, size 15 or 16 needles are suitable.

2 Care should be taken to see that 2 cm³ syringes and hypodermic needles both have 'luer' fittings; 'record' fitting needles are not suitable for 'luer' syringes.

*A useful book is: Rogers, M. (1970) *Gas syringe experiments*, Heinemann.

Rubber caps for 100 cm³ syringe

Packets of 1 dozen caps (Nuffield Physics catalogue item NF 148/2)	Philip Harris Ltd, 63 Ludgate Hill, Birmingham.	About 40*p* per packet.

Silicone rubber for self-sealing cap for hypodermic needle

The most convenient source appears to be silicone rubber bungs; from: (A suitable size is E5, quality T.C. 156/1)

ESCO (Rubber) Ltd,
Walshingham House,
35 Seething Lane,
London EC3

2   Apparatus to demonstrate the principle of the mass spectrometer (Topic 4)

Several different models which demonstrate this principle can be constructed. One of them is described here.

The model consists of a tray in which there are three compartments, as shown in figure A.1. Ball bearings of different masses are rolled down a glass tube into the tray and are then deflected by a magnet. Adjustment of the slope of the glass tube and of the position of the magnet enable the ball bearings to be deflected into the three compartments according to their masses.

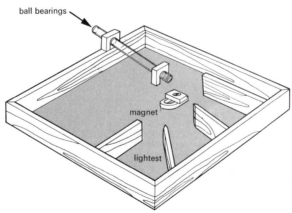

Figure A.1
Model to show the principle of the mass spectrometer.

Materials required

1 Eclipse horseshoe magnet – width across limbs 44 mm, thickness 10 mm. (Griffin and George No. L71–490) *or similar.*
Ball bearings diameters 11 mm, 6 mm, and 4 mm.
Wood for tray.
Glass tube about 30 cm long, internal diameter greater than 11 mm.

Construction

The dimensions of the tray are not critical. Nominal dimensions are given in figure A.2.

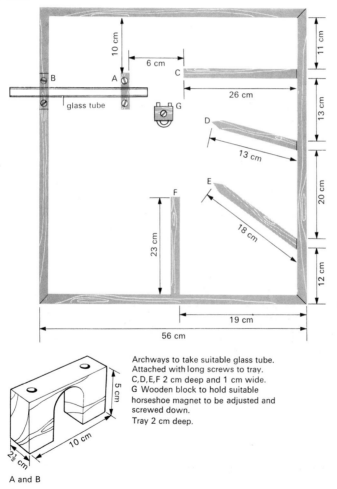

Archways to take suitable glass tube.
Attached with long screws to tray.
C,D,E,F 2 cm deep and 1 cm wide.
G Wooden block to hold suitable horseshoe magnet to be adjusted and screwed down.
Tray 2 cm deep.

A and B

Figure A.2
Details of the model.

Adjustment will be required, moving the magnet and tube, before the model works satisfactorily. During this adjustment, the ball bearings that stick to the magnet will become magnetized and should be dropped several times to demagnetize them.

3 Use of valves to find ionization energy (Topic 4)

The valve base, terminals and potentiometer should be mounted in such a way that the circuit can be readily seen.

Materials required

From Radiospares, 4–8 Maple Street, London W1.

4 red 4 mm sockets

4 black 4 mm sockets

2 yellow 4 mm sockets

1 wire wound volume control 250Ω, linear, 3 W

1 octal base

Suitably robust container such as a polythene food box, about 25×10 cm. Valve, either GT3 (price about 82*p*) or 884 (price about 55*p*); both obtainable from Z and I Aero Services Limited, 44 Westbourne Grove, London W2.

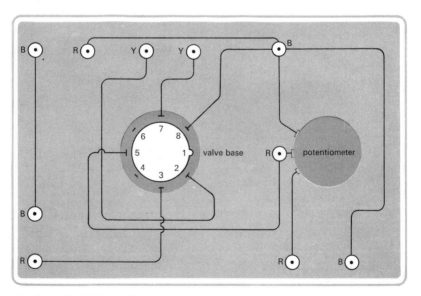

sockets R = red B = black Y = yellow

Figure A.3

Construction

The components should be mounted as shown in figure A.3.

The circuit should be shown on the lid of the box as in figure A.4.

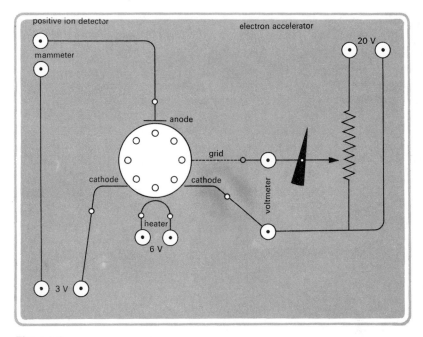

Figure A.4

4 Direct vision spectroscope (Topics 4 and 6)

The spectroscope can be made out of any hollow tube of a suitable size. One end of the tube is sealed except for a narrow slit, and a replica grating put over the other end.

Materials required

Tube about 12 cm long and 1–2 cm in diameter is suitable. The wider the tube, the longer the slit can be made and the deeper the spectrum. Dexion Speed-Frame electrical conduit or an empty thermometer or cigar case is suitable. Kitchen aluminium foil.

Replica diffraction grating. Sheets, 25×20 cm are available from Proops Brothers Limited, 52 Tottenham Court Road, London W1, price about £1.

Construction

The inside of the tube should be sprayed matt black to prevent internal reflections.

One end is covered with kitchen aluminium foil in which a slit is cut with a razor blade, and the other end is similarly covered and a 5 mm hole drilled in the foil. A piece of grating replica is stuck over the hole, so that the slit and the grating lines are parallel. The direct slit will be seen through the grating with the first order spectrum on one side.

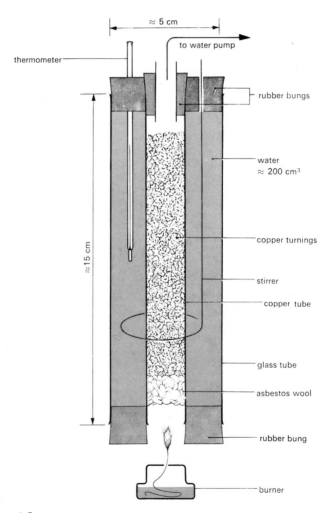

Figure A.5

5 Heat of combustion apparatus (Topic 7)

A suitable apparatus, devised by R. J. Sutcliffe, for determining the heat of combustion of alcohols can be made by mounting a copper tube containing copper turnings inside a glass tube containing water, as shown in figure A.5.

6 Bragg phase difference simulator (Topic 8)

The apparatus is shown in figures A.6, A.7, A.8, and A.9.

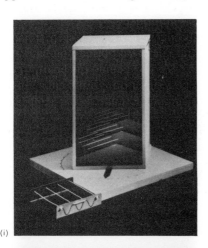

(i)

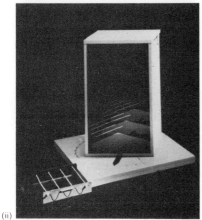

(ii)

Figure A.6
Optical analogue of Bragg X-ray diffraction.
i Reflected wave-trains out of phase.
ii Reflected wave-trains in phase.

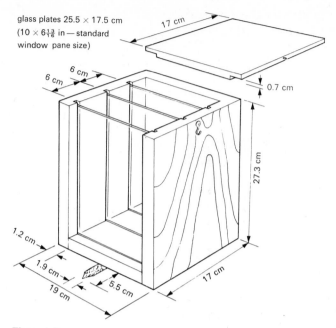

glass plates 25.5 × 17.5 cm
(10 × 6⅞ in — standard
window pane size)

17 cm

0.7 cm

6 cm

6 cm

27.3 cm

1.2 cm

1.9 cm

5.5 cm

19 cm

17 cm

Figure A.7
Box and lid.

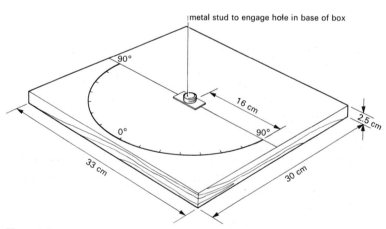

metal stud to engage hole in base of box

90°

16 cm

2.5 cm

0°

90°

33 cm

30 cm

Figure A.8
Base.

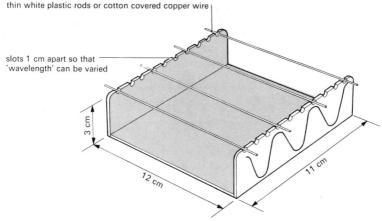

thin white plastic rods or cotton covered copper wire

slots 1 cm apart so that 'wavelength' can be varied

3 cm

12 cm

11 cm

Figure A.9
'Wave-train'.

The crystal is represented by three sheets of window pane glass (the glass plates shown in figure A.7 are a standard window pane size) supported inside a wooden box. The interior of the box should be blackened with a matt finish paint, to absorb light; and there should be a lid to the box. The box is on a pivot, and has a pointer which moves over a circular scale marked in 10s of degrees. A separate 10° segment marked in 1° divisions is a convenient method of reading to 1°.

The representation of the X-ray beam consists of a series of pieces of thick white cotton, very thin white string, cotton covered copper wire, or thin white plastic rods (as shown in figure A.9) supported on a metal former and equally spaced, and may be considered as wave crests. The side of the string nearest to the box is illuminated strongly by a desk lamp.

In the sheets of glass, multiple reflections of the threads may be seen. In most orientations of the 'crystal' this produces a jumble of images, but as the box is rotated the images of the threads move and coincide. The position of coincidence corresponds with maximum reinforcement, and at this point $2d \sin \theta = n\lambda$.

The angle θ is measured from the scale, λ is measured directly with a ruler, and from the equation d is calculated.

The box is then further rotated and a second order coincidence obtained. It is usually necessary to move the 'X-ray beam' over to one side in order to keep it in front of the glass. The beam must, however, always arrive in a direction parallel to the $\theta = 90°$ direction.

A convenient separation of the sheets of glass is 6 cm between adjacent sheets. A convenient wavelength is 3 cm. It is instructive to use a second wavelength, and 2 cm is appropriate.

7   Crossed polaroid assembly (Topic 8)
The assembly is shown in figure A.10.

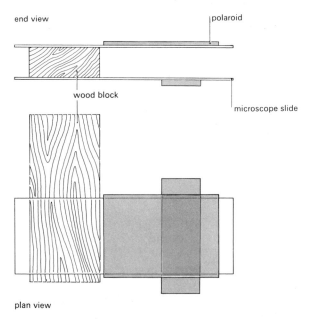

Figure A.10
Crossed polaroid assembly.

The microscope slides may conveniently be attached to the wood strip by adhesive tape. The lower piece of polaroid is attached by pieces of adhesive tape *at the edges*. (The tape must *not* overlap the central area of the polaroid, because it produces the type of optical effect which the experiment sets out to observe.) The assembly is mounted in a clampstand so that the microscope slides are horizontal, and a 3-volt torch bulb and holder are placed beneath the slides. With the light on, the upper piece of polaroid is adjusted until the light bulb, when viewed through it, appears to be extinguished. The upper polaroid is then taped in this position, being fixed at the sides or end only. Polaroid sheet is available from laboratory suppliers.

8 Polarimeter (Topic 13)

A polarimeter of the type shown in figure 13.5b may be obtained from Philip Harris Limited, 63 Ludgate Hill, Birmingham (price, about £5). A home-made version, devised by E. M. Somekh, is described in the *School Science Review* (1964), 158, 163. Another version of a home-made instrument, devised by K. Fraser, is shown in figures A.11 and A.12.

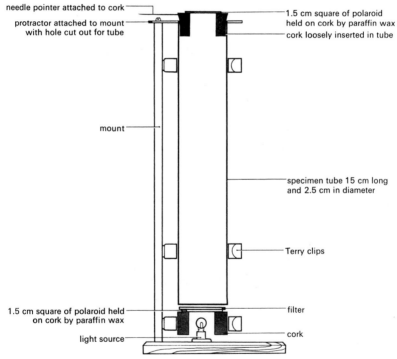

Figure A.11
A home-made polarimeter, side view.

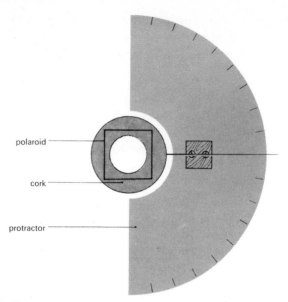

Figure A.12
A home-made polarimeter, viewed from above.

The analyser is mounted on the cork which is loosely inserted into the specimen tube. The polarizer is mounted below the specimen tube.

Materials required
2 pieces of polaroid, 1.5 cm square
Specimen tube, 150×25 mm
3 Terry clips
Wood for mount, $18 \times 2.5 \times 2.5$ cm
Protractor, $0-180°$, 10 cm diameter
Sewing needle for pointer
2 corks
Paraffin wax
6 V bulb and holder

Note. If the version shown above is used in experiment 13.5b the volumes of solution will need to be increased because of the larger capacity of the specimen tube used.

9 Colorimeter (Topics 14 and 16)
A colorimeter affords a quick and easy method of determining the concentrations of solutions which are coloured. Basically it consists of the components shown in figure A.13. (For details of a more sophisticated colorimeter, see Tetlow, K. S. *School Science Review* (1970) **51**, 176, 637. For a simpler version see *S.S.R.* (1970) **51**, 177, 924.)

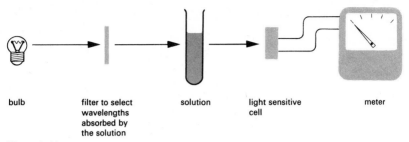

Figure A.13
Colorimeter.

The light sensitive cell may either be a selenium cell which produces an e.m.f. proportional to the intensity of the light falling on it, or may be a cadmium sulphide cell, the electrical resistance of which is proportional to the intensity of the light falling on it. Either way the meter reading gives an indication of the intensity of light emerging from the solution.

The connection between the intensity of light emerging from the solution and the molarity of the absorbing species in the solution is given by:

$$\lg\left(\frac{I_0}{I}\right) = \lg\left(\frac{m_0}{m}\right) = \varepsilon l M$$

or $\lg\left(\dfrac{I_0}{I}\right) = \lg\left(\dfrac{m_0}{m}\right) \propto \mathbf{M}$

where I_0 = intensity of incident, *monochromatic* light giving meter reading m_0
       I = intensity of emergent light giving meter reading m
       M = molarity of solution
       l = path length of light through solution
       ε = molecular extinction coefficient

The colorimeter is normally used by first adjusting the meter reading to maximum for I_0 by inserting a tube of pure solvent, and then altering the intensity of light by adjusting a shutter placed between the bulb and the cell (or by altering the current flowing through the bulb). The tube of solution is then put into the colorimeter and a meter reading proportional to I determined. It is *most important* that this procedure is adopted for every reading if possible, unless the electricity supply is very stable.

Unfortunately, the law quoted above is not obeyed accurately by all solutions. Also the meter reading may not be accurately proportional to the intensity of light, so the instrument must be calibrated before use. This will also indicate for what range of concentration of a particular substance the colorimeter can be used.

For permanganate solutions in a particular colorimeter the curve shown in figure A.14 was obtained.

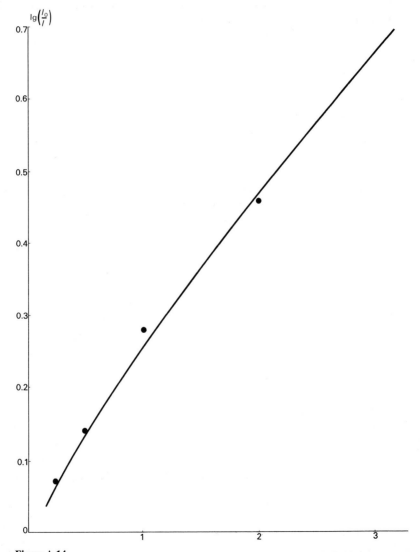

Figure A.14
A calibration curve for the colorimeter.

If a calibration curve is being used, it is easier to plot $\left(\dfrac{I}{I_0}\right) = \left(\dfrac{m}{m_0}\right)$ than $\lg\left(\dfrac{I_0}{I}\right) = \lg\left(\dfrac{m_0}{m}\right)$, as m_0 is normally 1, 10, or 100 or some number which is a useful number to have as a denominator.

Such a graph (figure A.15) indicates that the colorimeter will not measure concentrations accurately above about 0.0003M.

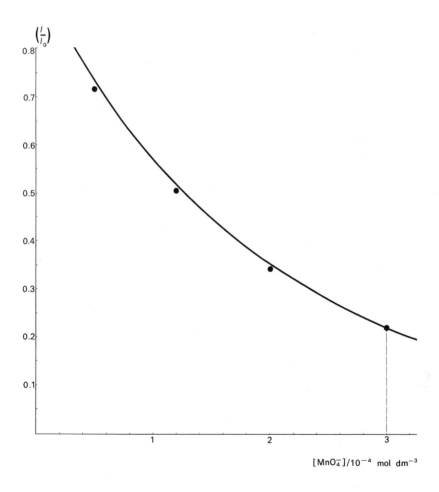

Figure A.15
An alternative form of calibration curve.

Before these curves can be constructed, the most suitable filter must be chosen. This should select that band of wavelengths of light which are most strongly absorbed by the solution. This is satisfied by the filter which gives the lowest value of $\left(\dfrac{I}{I_0}\right)$ for a particular solution.

Procedure

It is convenient to supply students with a calibration curve for a particular filter, colorimeter, and solution. They will require a test-tube for the pure solvent and an optically matched one for the solution. To take a reading the meter is adjusted to maximum with pure solvent in place, and then the reading is obtained with solution in place. It is best then to check again with solvent in place and obtain a meter reading with solution in place once again. The value of $\left(\dfrac{I}{I_0}\right)$ so obtained can then be turned into a molarity using the calibration chart.

Construction of a colorimeter

Several colorimeters are now on the market, the cheapest being that made by Evans Electroselenium Limited, consisting of the essential parts of a more expensive colorimeter. A colorimeter can be constructed by mounting a selenium cell, which produces an e.m.f. proportional to the illumination falling on the cell, together with a bulb on a suitably drilled block which contains a hole for a test-tube and a slit for the filter, as shown in figure A.16.

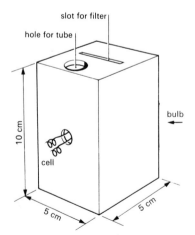

Figure A.16

There has to be some means of varying the intensity of light reaching the cell and an iris from an old camera mounted in front of the bulb is ideal. If this is not available, the intensity of the bulb can be altered by wiring it in series with a variable resistance, but as this alters the spectrum produced by the bulb the shutter method is preferable. The 17×43.5 mm rectangular selenium cell made by EEL is suitable for such a colorimeter and a meter with 50 μA full scale deflection will also be required. One with both linear and log scales (such as that made by British Physical Laboratories) is particularly useful.

A cadmium sulphide cell, such as the ORP 12*, the resistance of which changes with the illumination falling on the cell, can be used instead of the selenium cell. It has to be wired in series with a 6 V supply and a 1 mA meter so that its resistance can be monitored.

Ideally a set of matched sample tubes with identical optical properties should be used, but a set with sufficiently similar properties can be selected from ordinary test-tubes.

Materials required

Wood for colorimeter, $5 \times 5 \times 10$ cm
Bright spectrum filter set, gelatin, unmounted, 2×2 in. (Ilford)
Each filter when mounted between microscope slides provides 4 sets of filters

either   Selenium cell such as 17×43.5 mm cell made by Evans Electroselenium Limited, Halstead, Essex
50μA full scale deflection meter such as that with both linear and log scales made by British Physical Laboratories, Radlett, Herts

or   Cadmium sulphide cell ORP 12, obtainable cheaply from various sources (advertisement pages of *Practical electronics*)
1 mA full scale deflection meter
Source of 6 V d.c. (0.3–0.4 A)†

either   Suitable iris from old camera

or   From Radiospares, 4–8 Maple Street, London W1
1 wire-wound volume control, 100Ω linear 3W ⎫
1 wire-wound volume control, 500Ω linear 1W ⎬ nominal values
Carbon resistance $\approx 70\Omega$ ⎭
1 lampholder (clip-on M.E.S.)
1 bulb 6.5 V 0.3 A (M.E.S.)
4 mm plugs and sockets

*The response of the ORP 12 is inferior to some other photoresistors but it is adequate for use in this course and is cheaper.

†Accumulators are ideal for this. If mains current is used and fluctuation occurs, a cheaply constructed voltage stabilizer (6–12V, 1A) is described in *Practical electronics*, Volume 5, No. 7, page 521 (July, 1969).

These can be conveniently mounted in a polythene box, the circuit being clearly shown on the lid. If a cadmium sulphide cell is being used, its circuit can also be included in the box as shown in figure A.17.

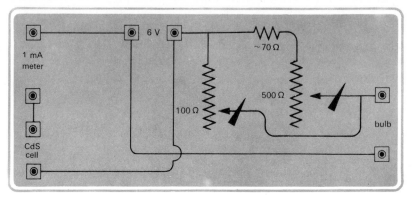

Figure A.17

The 100Ω potentiometer gives a coarse adjustment, and the 500Ω potentiometer a fine adjustment of the bulb intensity.

10 Potentiometers for e.m.f. measurements (Topic 15)
A potentiometer is used to measure a potential difference accurately. The basic circuit is shown in figure A.18.

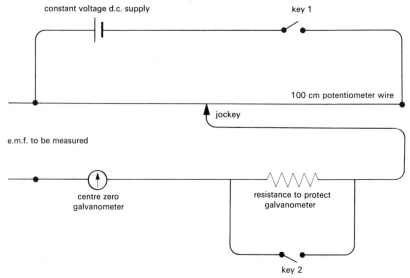

Figure A.18

The potentiometer wire is calibrated before use by noting the length of wire corresponding to a known e.m.f. from a standard cell. If a standard cell such as the Weston is not available, a mercury cell is a possible substitute. The Mallory RM–1 or RM–1R is suitable for this.

There are several ways in which suitable apparatus can be obtained.

a   Standard Physics apparatus can be bought, or borrowed from the Physics Department. For each potentiometer the following materials will be needed:

2 keys for making and breaking the circuit
1 straight wire potentiometer, 100 cm^3, 0.05Ω cm^{-1} with jockey
1 centre zero galvanometer (sensitivity at least 10 μA per division)
1 protective resistance for galvanometer, approximately 1000Ω
1 2 V accumulator
Leads
Standard cell, Weston or mercury battery

b   Many of the components required for a straight wire potentiometer can be made quite cheaply.
100 cm^3 Constantan wire (S.W.G. 28) can be mounted on a suitable length of wood together with a metre rule.
A meter, switches and terminals can be mounted in a polythene box on the lid of which the circuit is clearly shown as in figure A.19.

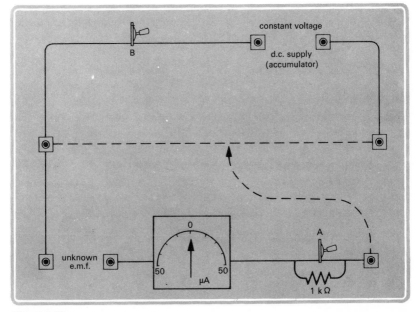

Figure A.19

The components required are:

From Radiospares, 4–8 Maple Street, London W1
7 4 mm sockets
1 kΩ $\frac{1}{2}$W carbon resistor
1 SPST biased switch (A) ⎱
1 SPST switch (B)   ⎰ not essential

and   4 oz (100 m) Constantan wire (S.W.G. 28)
1 metre rule

or   metre scale paper

also   1 polythene box
Wood for mounting potentiometer wire
1 jockey made from small screwdriver

and   50–0–50 μA galvanometer (Shinohara $1\frac{21}{32}$ in square from G. W. Smith, 3 Lisle Street,
London WC2, is a suitable cheap meter (about £1.50)
A 2 V accumulator and standard cell are also required

c   A potentiometer using a helical potentiometer instead of a straight wire can be used. Details are given in the Nuffield Advanced Chemistry *Apparatus and materials list* obtainable from the Association for Science Education, College Lane, Hatfield, Herts.

11   Hydrogen electrode (Topic 15)
A number of quite simple versions of this electrode can easily be constructed. Three possible designs are shown in figure A.20.

Expense is saved if copper wire is soldered to a short length of platinum wire (use zinc chloride/hydrochloric acid flux and ordinary soft solder).

The platinum wire must be covered with electrolytically-deposited platinum over the lower inch or so. This is done by using another platinum wire as a second electrode, and immersing both in a 3 per cent solution of platinum(IV) chloride containing a small proportion (approximately 0.05 per cent) of lead acetate, connecting the electrodes to a d.c. supply (about 4 V), and reversing the current every half minute for 15 minutes. The electrodes are then washed with distilled water, and occluded chlorine is removed by placing the electrodes in sodium acetate solution (approximately M) and again passing current for 15 minutes, reversing every half minute. The electrodes are then washed with distilled water and stored in distilled water when not in use.

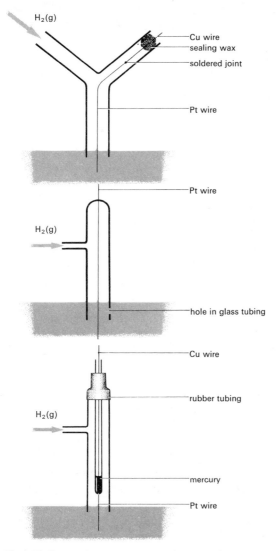

H₂(g)
Cu wire
sealing wax
soldered joint
Pt wire
Pt wire
H₂(g)
hole in glass tubing
Cu wire
rubber tubing
H₂(g)
mercury
Pt wire

Figure A.20
Types of hydrogen electrode.

A cylinder is most convenient supply of hydrogen gas but must be fitted with a reducing valve, in order to provide a very slow stream of gas. Alternatively, a filter flask can be used as an aspirator (see figure A.21). This can be filled from a cylinder or from a suitable gas generating apparatus (in this case put dilute permanganate solution and a little dilute acid in the aspirator flask to remove impurities liable to poison the platinum electrode).

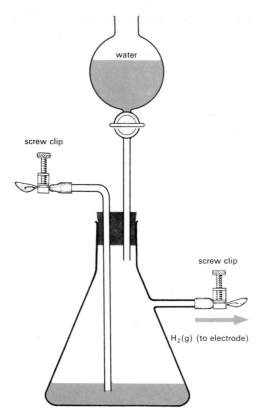

Figure A.21
An aspirator for a hydrogen electrode.

12 Antimony electrode (Topic 15)

This electrode can be used for the determination of the pH of a solution. It is an antimony/antimony oxide electrode and the oxide is in equilibrium with antimony ions and hydroxide ions in solution:

$$Sb_2O_3(s) + 3H_2O(l) \rightleftharpoons 2Sb^{3+}(aq) + 6OH^-(aq) \tag{1}$$

Thus the potential of the electrode is dependent upon the concentration of hydroxide ions in solution, amongst other things. As $[OH^-(aq)] = \dfrac{K_w}{[H^+(aq)]}$ in aqueous solution, this means that the potential must be dependent upon the hydrogen ion concentration of the solution.

As $E_{Sb} = E_{Sb}^{\ominus} + \dfrac{RT}{zF} \ln [Sb^{3+}(aq)]$ and as $[Sb^{3+}(aq)]$ can be related to $[H^+(aq)]$

by means of the equilibrium constant, K_1, for equation (1) and the ionic product of water,

$$E_{Sb} = E'_{Sb} + \frac{RT}{F} \ln [H^+(aq)]$$

where E'_{Sb} is a constant involving $E^\ominus_{Sb}$, K_1 and K_w.

Construction and operation of the electrode

2 or 3 cm of antimony stick is cleaned with emery paper and a copper wire soldered on to one end. It can be mounted in a glass tube and sealed in with sealing wax. The electrode is then cleaned again and kept under water.

To calibrate the electrode, the cell:

$$Sb(s), Sb_2O_3(s) \, | \, [H^+(aq)] \, \vdots \, Cu^2(aq) \, | \, Cu(s)$$

is set up with a standard copper electrode and the e.m.f. determined with the antimony electrode immersed in buffer solutions of known pH. (A new salt bridge should be used with each new solution.) The calibration curve thus obtained will then enable the electrode to be used to determine the pH of solutions, the pH of which is not known. The cell is relatively insensitive to small temperature changes. A typical calibration curve is shown in figure A.22.

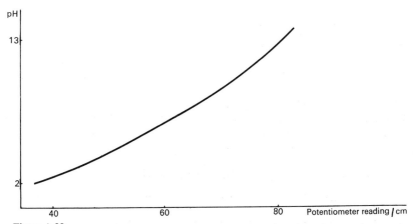

Figure A.22
A calibration curve for use with an antimony electrode.

Tablets for making solutions of pH 4, 7, and 9 can be purchased from British Drug Houses Limited or their 'Universal buffer mixture' can be used to make the whole range of solutions required. An alternative is to make up all the solutions required from available chemicals. For the pH values needed, the following mixtures (taken from Findlay's *Practical physical chemistry*, 8th edition) can be made up from substances which are commonly found in a chemical store.

pH

 2   5.3 cm^3 0.2M HCl + 25 cm^3 0.2M KCl; dilute to 100 cm^3
 4   41.0 cm^3 0.2M acetic acid + 9.0 cm^3 0.2M sodium acetate
 7   12.0 cm^3 0.05M borax + 188.0 cm^3 of a solution containing 12.4 g boric acid and 2.93 g sodium chloride per dm^3
 9   40.0 cm^3 0.05M borax + 10 cm^3 of above H_3BO_3/NaCl solution
10   Equal volumes of 0.025M NaHCO$_3$ and 0.025M Na$_2$CO$_3$.

13   Smooth platinum electrodes (Topic 15)
(These are often available from Hofmann and similar voltameters.)

   a   Fold a 1 cm^2 piece of platinum foil over a 5 cm length of platinum wire, heat to redness, and squeeze with pliers. Repeat about six times to weld together and then open out the foil.

   b   Solder a length of copper wire to the platinum wire (ZnCl$_2$ + HCl flux and soft solder). Fuse the platinum wire into a glass tube. Seal the wire into the other end of the glass tube with sealing wax or plaster of Paris.

Materials

Mineral specimens
These are obtainable from:
   R. F. D. Parkinson & Co. Ltd,
   Doulting, Shepton Mallet, Somerset.

   Gregory, Bottley and Co.,
   30 Old Church Street, London SW3.

   The Geological Laboratories,
   168 Moss Lane East, Manchester 15.

The materials required for the course are:

Demonstration specimens. Recommended size, 5 × 5 cm.

Halite (sodium chloride)
Calcite, var. Iceland Spar
Calcite, Minsterley, Shropshire
Fluorite, cleaved octahedron
Fluorite (Weardale), cubic crystals
Graphite
Quartz, var. Rock Crystal, Bristol Diamonds
Quartz, var. Rock Crystal (Brazil), single crystal
Gypsum, var. Selenite (Trowbridge)
Anhydrite
Mica, var. Muscovite
Mica, var. Biotite

Total cost: about £5

For class use

Halite (sodium chloride)	1 kg
Calcite, Minsterley	1 kg
Calcite, var. Iceland Spar	One 2.5 × 2.5 cm crystal
Fluorite (Menheniot)	250 g
Gypsum, var. Selenite, for cleaving	1 kg
Mica, var. Muscovite	250 g

Total cost: about £2.50

Crystalline sodium chloride (from a melt), for demonstration specimen and for class cleaving.

Philip Harris Ltd, 63 Ludgate Hill, Birmingham.

250 g required. Price: about £3.50 per 250 g.

Reagents

Unless otherwise specified in this Guide, common bench reagents are assumed to be about 2M. In the preparation of the somewhat numerous other solutions required much time is saved if a stock of concentrated solutions of fairly accurately known concentration is maintained. These can then be suitably diluted when needed. For solutions whose concentrations need to be known accurately, the concentrated volumetric solutions which can be purchased from chemical suppliers are very convenient but rather expensive.

Appendix 3

Models and other visual aids

Structure models for the course

In any serious attempt to assist students to understand the three-dimensional structures of substances the need to use models should not need justifying. Any teacher who attempts to construct structure models will rapidly find that a detailed understanding of the spatial relationships between atoms is necessary in order to complete them successfully. Therefore students will best learn about structures if they construct models for themselves, and a 'model workshop' is likely to be an annual feature of this chemistry course.

Given that the objective is to construct models each year it is necessary either to have component parts (figure M.1) from which models can be readily assembled when required and then demounted for the following year, or to construct models from cheap materials which can be regarded as expendable and perhaps sold to their constructors to defray the cost. There are, however, a number of models that the teacher will wish to have in permanent form because of their complexity or because their main teaching use will be for display and discussion rather than for consideration of how the structure is built up.

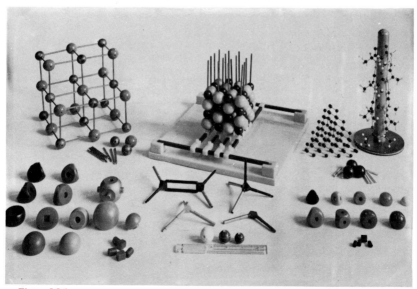

Figure M.1

Examples of structure models. *Back row:* demountable ball-and-spoke (*Crystal structures*); space-filling, ionic structures (*Catalin*); miniature models (*Beevers*); Linell ball-and-spring (*Gallenkamp*). *Front row:* Stuart-type organic set (*Griffin and George*); Dreiding stereomodels (*Rinco, USA*); space-filling covalent set (*Catalin*).

For this course a number of crystal structure models of the ball-and-spoke type are required. *Caesium chloride, diamond, graphite, ice, sodium chloride (rock salt), zinc sulphide (zinc blende)* can all be obtained from firms such as Beevers' Miniature Models or Crystal Structures Limited. Beevers' miniature models are supplied as assembled models or parts, but it is probably best to purchase them assembled and treat them as permanent models; an *α-helix*, based on alanine, and *quartz* structures are also available from Beevers. The models are beautifully made, robust, and economical. Crystal Structures supply ball-and-spoke demountable models (Z series) and models of *calcium carbonate (calcite)* and *calcium fluoride (fluorspar, fluorite)* can also be obtained from them. The models are made to a generous scale, readily assembled and dis-assembled, robust, and economical.

The models listed so far can be purchased for under £25.

Also required are space-filling structure models to illustrate *hexagonal* and *cubic close packing, body-centred cubic packing*, and the ionic structures of *sodium chloride, caesium chloride*, and *aluminium chloride*. All of these structure models can be purchased in packs of twelve model atoms at about £0.40 a packet. The necessary components (standard base, 2 side rails and a minimum of 4 cross rails, 16 rods and 32 model ions of each type but rather fewer aluminium) can be purchased for under £15.

It is helpful if models of simple inorganic and organic molecules can be constructed. The necessary components can be of the ball-and-spoke type or space-filling. The models suggested in the course are $BeCl_2$, BF_3, $SiCl_4$, SiH_4, NH_3, P_4, P_4O_{10}, PCl_3, PCl_5, PH_3, H_2O, S_8, H_2S, HF, Cl_2, HCl, and CH_4, $CHCl_3$, CCl_4, HCN, CO_2, $CO(NH_2)_2$, C_2H_2, C_2H_4, C_2H_6, C_2H_5Cl, C_4H_9Cl, C_4H_9Br, C_4H_9I. Very suitable ball-and-spring components are the Linnell type which can be purchased in packs of twelve model atoms at about £0.40 a packet. Linnell components can also be used to make some of the crystal structures referred to above, for example caesium chloride, diamond, graphite, ice, sodium chloride, zinc sulphide, and calcium fluoride. The following quantities would be required to make suitably sized models:

Caesium chloride:
8 metal ions (yellow, 8 coordinated)
27 chloride ions (green, 8 coordinated)
96 springs, about 6 cm length

Diamond:
30 carbon atoms (black, tetrahedral)
30 springs, about 6 cm. length. This is sufficient for four atoms along each side of a tetrahedron; a more appropriate size requires 60 atoms and 60 springs.
Six membered rings are made in the 'chair' form and then assembled into a lattice with 4 coordination.

Graphite:
48 carbon atoms (black, 5 hole trigonal bipyramid)
48 springs, about 6 cm length
8 spokes, about 12 cm length

Ice:
24 metal ions (brown, 8 coordinated)
24 sulphur atoms (yellow, 4 coordinated)
48 springs, about 6 cm length
These components are used to produce a wurtzite, zinc sulphide, structure which is one of the structures adopted by ice.
If six-membered rings of alternate zinc and sulphur atoms are made in the 'boat' form, and then assembled into a lattice, with 4,4 coordination, the wurtzite structure results.

Sodium chloride:
32 metal ions (yellow, 6 coordinated)
32 chloride ions (green, 6 coordinated)
144 springs, about 6 cm length.

Zinc sulphide:
(zinc blende)
15 metal ions (brown, 8 coordinated)
15 sulphur atoms (yellow, 4 coordinated)
30 springs, about 6 cm length.

Calcium fluoride:
(fluorite)
15 metal ions (brown, 8 coordinated
12 halogen ions (green, 8 coordinated)
54 springs about 6 cm length.

In constructing these crystal structure models the illustrations in the Overhead Projector Printed Originals should serve as a helpful guide. Sufficient Linnell components for the inorganic molecule models and the crystal structure models would cost under £25 but to obtain the complete set of crystal structure models about a further £10 would have to be spent.

The models of inorganic molecules could also be obtained by the purchase of the necessary standards packs and components of Catalin Covalent Models.

By the use of rubber pegs as supplied, space-filling models can be rapidly assembled and 10 cm steel connecting pins are also available to convert the models to the ball-and-spoke type. The particular merit of these models is the wide range of accurately sized and shaped atom models to represent for example the individual halogens or trivalent, phosphate, and pentavalent phosphorus atoms. A suitable range of components would cost under £15.

The learning of *organic chemistry* can be made more effective if students are asked to construct models as each new functional group is discussed. The shapes of organic isomers can have a great influence on their reactivity and students need to be made continually aware of the need to consider the shapes of molecules when discussing their reactivity and possible reaction pathways. For this aspect of the work to be effective large numbers of components need to be available so that each student can construct his own models. Models can be either space-filling or ball-and-spoke but if a large collection of one type is available it is desirable also to have a small set of the other type. A space-filling set of Stuart-type atomic models is available imported from Japan. The models are blow-moulded plastic shapes and an occasional item will be poorly finished but they are robust, readily assembled, and relatively very cheap with a set of 336 model atoms costing under £25. With a set of this size a complete class can participate in the construction of models rather than the activity being restricted to the teacher and a few students at a time. These models are used to illustrate Topics 9 and 13 in the *Students' Book*. Another inexpensive model set, *Scale atoms*, has also appeared recently.

As an alternative an excellent set of ball-and-spoke models has been developed by Louis Fieser from the original work of Dreiding and Bretscheneider. The tetrahedral carbon models have two plastic rods and two aluminium tubes to represent bonds and robust models can be assembled with great speed by inserting rods into tubes. Furthermore, since carbon-carbon double bonds are moulded as a single unit containing the two carbon atoms it is easy to select the appropriate components when assembling a model. The whole system is ingenious and produces elegant models at reasonable cost. (The present limitation is that the models have to be purchased direct from the USA but this is readily done through a bank by means of a dollar draft.)

Teachers will appreciate that the models described above are representative of what is available at the time of writing and catalogues should be consulted for up to date information on the above and other kits.

A number of the structure models listed above can be made more cheaply from wooden beads or polystyrene spheres if adequate time is available to organize the materials. Students should be able to construct the smaller models from polystyrene spheres at a sufficiently low cost for the materials to be discarded or the models sold to their constructors. A guide to model making is given in the Nuffield Chemistry *Handbook for teachers*. Chapter 14 has information on constructional techniques, supplies of materials, and references to the numerous articles published in the *School Science Review* and elsewhere giving constructional details. For the following structure models specific details are given in Nuffield Chemistry *Collected Experiments: space filling sodium chloride and caesium chloride* (E13.10, page 252), *sulphur molecule* (E13.15, page 261), *diamond and graphite* (E13.17, page 263), *ice* (E13.19, page 264).

One model that is not described elsewhere is a space-filling model of *aluminium chloride*. This is a layer structure with the sequence Cl–Al–Cl, Cl–Al–Cl, and so on. The chlorine layers have the hexagonal close packed arrangement with aluminium atoms occupying two-thirds of the octahedral holes.

Make a layer of polystyrene sphere chlorine atoms (4 cm diameter) in h.c.p. arrangement, attaching them together by means of drops of *saturated* solution of polystyrene in pentyl acetate (amyl acetate). Insert polystyrene sphere (1 cm diameter) aluminium atoms in two-thirds of the holes, as shown in the following diagram.

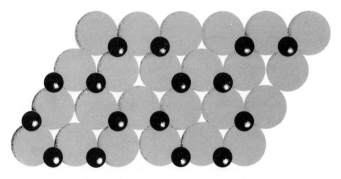

Figure M.2
Part of the aluminium chloride structure.

Arrange another layer of chlorine atoms over the first, in hexagonal close packed arrangement. The chlorine spheres should be coloured green.

For further information on the models mentioned in this section write to:

Beevers' Miniature Models	Dr Beevers, Chemistry Department, Edinburgh University, West Mains Road, Edinburgh 9, Scotland.
Crystal structures	Crystal Structures Ltd, Bottisham, Cambridge.
Linnell type components	Gallenkamp Ltd, Technico House, Christopher Street, London EC2.
Catalin Covalent Models Catalin Ionic Models	Catalin Ltd, Waltham Abbey, Essex.
Stuart-type atomic models	Griffin and George Ltd, Ealing Road, Alperton, Wembley, Middlesex.
Dreiding plastic stereomodels	Rinco Instrument Co. Inc., 503 S. Prairie Street, Greenville, Illinois 62246, USA.
Polystyrene spheres	Elford Plastics Ltd, Brookfield Works, Wood Street, Elland, Yorkshire.
Scale atoms	Heinemann Educational Books Ltd, 48 Charles Street, London W1X 8AH.

Overhead projection originals

The overhead projector is a modern form of the diascope, that is, it projects still pictures onto a screen by passing light through a transparency. The transparencies used are up to 25 cm square and a quartz-halogen light source in conjunction with a Fresnel lens produces a 1.5 metre square or larger picture suitable for viewing in normal lighting. The projector is used 2.5 to 3.5 metres from the screen and it is possible for the teacher to sit alongside to point out detail on the transparency. There is no need for the teacher to observe the screen and he can maintain continuous, and informal, contact with the class. The projector can also be fitted with a 15 metre roll of transparent film on a special roller attachment to serve as a substitute for the chalkboard.

A full account of the use of the overhead projector is given in the booklet *The overhead projector* by A. Vincent (EFVA) and to assist in the choice of a particular model teachers will find it helpful to consult *A survey of overhead projectors* by A. H. Crocker (EFVA).

For this course a set of *Overhead projection originals* has been prepared. They can be used in three ways: as originals from which transparencies can be prepared for use on the overhead projector, as originals from which duplicate copies can be produced for distribution to students, or as small scale wallcharts to be pinned up in the classroom.

Figure M.3
An overhead projector in use. *Projector by 3M Company*

To prepare transparencies from printed originals, methods such as a photographic diffusion process or a thermal process are available. The different types of transparency which can be prepared by the thermal process make it particularly suitable for the chemistry printed originals and the method is very simple. The infra-red copier machine is first warmed up by passing a plain sheet of paper through two or three times, the temperature control dial is then set to give an appropriate machine speed, and finally a sheet of specially coated acetate film is placed on top of the printed original and passed through the copier. The transparency is produced in a few seconds ready for use.

The 3M Company manufacture a variety of coated acetate films. Type 133 is a thin but very strong sheet on which a black image is produced by the thermal process. It is suitable for most of the chemistry printed originals. Type 127 is a thicker sheet useful for transparencies which are likely to be written on (see below) or frequently handled. If an original contains fine detail, or if typewritten information has been added, Type 125 is more suitable as the thermal process produces a fine line frosted image. There is also Type 888 available as a 'rainbow pack' from which transparencies can be prepared in five different colours. It is especially useful for the preparation of overlays (see below).

Transparencies have certain advantages over exclusive reliance on the chalk-board. For example tables of data can be readily presented for discussion, diagrams of complex apparatus are easily shown to students, and molecular structure models are available as accurate drawings. And all of these items are available for repeat presentations if required.

Transparencies can be used in a variety of ways. Thus the full content need not be revealed at once but portions may be covered with opaque material and exposed only when required. In other cases the content can be on a set of separate transparencies which in use are overlaid on each other in sequence. The transparencies can also be written on, or important portions highlighted with colour, by means of water-based fibre-tipped pens, and afterwards the transparency can be wiped clean for re-use.

The storage of transparencies so that they are readily available for use needs to be properly organized. If each transparency is fixed to a cardboard mounting frame (using a pressure sensitive tape) they can be numbered on the frame and stored in an ordinary cardboard box in numerical order. Kept in this way a collection of a hundred transparencies can be handled with ease.

Suppliers of the items mentioned are as follows:

EFVA booklets	Educational Foundation for Visual Aids, 33 Queen Anne Street, London W1 M OAL.
Transparency material	3M Company, 3M House, Wigmore Street, London W1.
'Lumocolor' projection pens	Matthews, Drew, and Shelbourne Ltd, 78 High Holborn, London WC1.

Film loops

A set of film loops is available to support the teaching of the course. Students nowadays are attuned to learning through visual media and a film loop can be a very effective learning aid. But to make its effect a film loop needs a suitable commentary and this is best done by the teacher developing his own from the commentary provided in the teaching notes with each film loop. In this way the commentary can be closely matched to the needs of the students. A suitable commentary makes some of the film loops prepared for the Nuffield O level Sample Scheme equally useful for the advanced course.

When showing the advanced chemistry film loops it is desirable to use a projector with a 'still picture' device operated from a remote control switch (figure M.4). This enables the film loop to be stopped for a discussion of picture content whenever required. The 'still picture' device is also most useful when students wish to raise questions about particular details in a film loop.

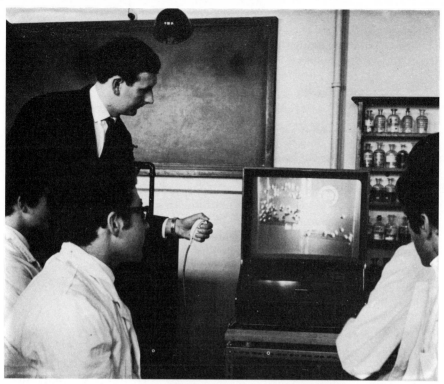

Figure M.4
A film loop projector in use in a laboratory. *Projector by Sound-Services Limited*

The following film loops have been prepared for the advanced chemistry course
in collaboration with World Wide Pictures. They are available in Standard 8.

	Section
Applications of the mass spectrometer	**4.3**
The Born-Haber cycle	**7.4**
Organic analysis by the mass spectrometer	**9.1**
Addition to carbon-carbon double bonds	**9.2**
Problems in the use of detergents	**9.5**
Applications of paper chromatography	**13.5**
The hydrolysis of bromoalkanes	**14.3**
Rate of reaction	**14.4**
Two-way paper chromatography	**18.3**
The manufacture of plastic articles	**18.6**
Testing of plastic film	**18.6**

Other film loops are available which have been found useful in the teaching of this course. They are mentioned under the appropriate topics and can be obtained from a film loop distributor such as Sound-Services Limited, Kingston Road, London SW19.

Display Boards

Teachers will find it useful when teaching this course to have available an energy level display board and a chemical specimens display board.

The *energy level display board* has its most important use in Topic 4 when electronic energy levels are first discussed. As well as being an effective method of displaying energy levels and the arrangement of electrons within those levels the board is also useful because an important concept is introduced by means of an interesting demonstration and thereby rendered more memorable for many students. The display board (figures M.5 and M.6) is readily constructed from plywood with wooden shelves to represent the energy levels. Beads of 2 cm diameter are used to represent electrons and are fitted over headless nails on the energy levels as required. A box fitted behind the board can be used to store a supply of beads.

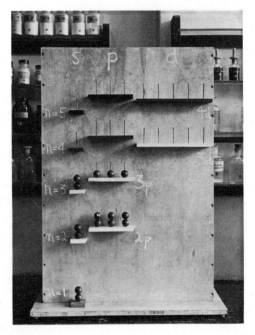

Figure M.5
An energy level display board.

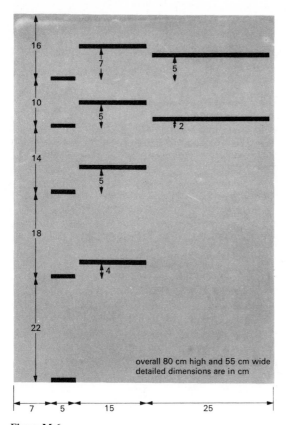

Figure M.6
Suitable dimensions for an energy level display board.

The *chemical specimens display board* is particularly useful when used to support the study of inorganic substances in relationship to the Periodic Table. For example during the study of Topic 2 the display board should be continuously present and carry (as far as possible) specimens of the chlorides and oxides of the second and third short period elements. The visible presence of the compounds is an effective reminder to the students of the change in properties that occurs across a period. Similarly in Topic 5 samples of the halogens and their compounds can be added to the board as they are investigated but when concluding the topic the specimens should be re-arranged to illustrate the oxidation numbers of the halogens in their compounds. Another good time to use the display board is during Topic 16 when d-block elements can be displayed in their different oxidation numbers: here, as elsewhere, students can be encouraged to build up the display using their own laboratory material. The display board (figures M.7 and M.8) should be constructed from a plastic surfaced plywood so that the face can be written on. Spring clips are used to hold specimen tubes which can be stored in a box behind the board. 'One inch size' plastic coated clips are available from The Lewis Spring Company Limited, Resilient Works, Redditch. Suitable specimen tubes are Ref. 3/H/3906 from Camlab (Glass) Limited, Milton Road, Cambridge.

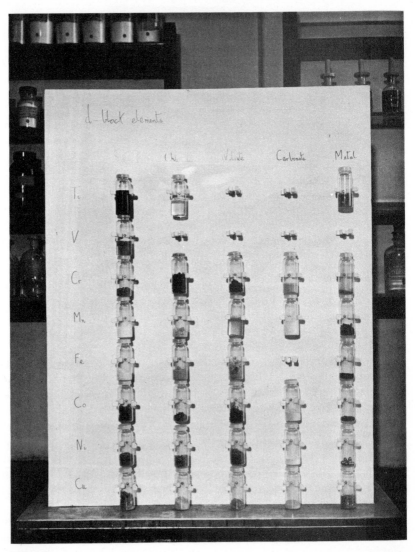

Figure M.7
A chemical specimen display board.

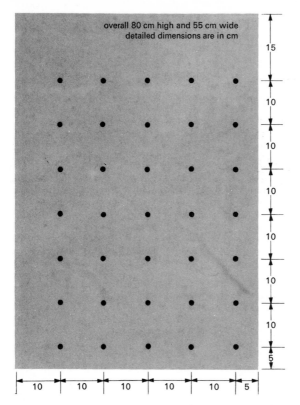

overall 80 cm high and 55 cm wide
detailed dimensions are in cm

Figure M.8
Suitable dimensions for a chemical specimens display board. Each dot represents a plastic coated clip.

Wallcharts
The questions of availability, storage, and protection of wallcharts is discussed in the special audio-visual aids issue of *Education in chemistry* for May 1968. Teachers should consult the article 'Wallcharts in chemistry teaching' by K. M. Harrison for further information.

A series of wallcharts is being prepared by the Royal Institute of Chemistry, 30 Russell Square, London WC1, in consultation with members of Head-quarters Team of the Chemistry project. The first three titles in the series will be:

'Structures of the hydrides'
'Indicators and pH'
'Redox potentials'

It is proposed that further charts will be produced in this series.

Appendix 4

Safety

A correct attitude to safety is of prime importance to the student; he must learn to work safely but without worry. A set of rules governing conduct in the laboratory is given below, and these should be rigidly enforced. The student must be taught to handle the apparatus and chemicals he uses safely and competently as a matter of habit rather than conscious thought. With the aid of a competent and conscientious teacher a student can readily slip into a safe system of working and, provided new hazards are pointed out to him as they arise in the normal course of his work, there should be no problems.

Laboratory regulations
1 *Never* taste unknown substances.
2 Always wear goggles and protective gloves when pouring corrosive liquids.
3 Never force glass rod or tubing into rubber bungs. Use glycerine as a lubricant and *always* protect the hands with a cloth.
4 Always carry out reactions involving poisonous gases in the fume cupboard.
5 Wear protective gloves when using compounds which are radioactive or harmful to the skin.
6 Do not throw refuse down the sink.
7 Broken glass must be thrown only in the special bin provided.
8 Ensure that all cuts are properly treated.
9 Foreign bodies in the eye – notify the teacher immediately.
10 Never use inflammable liquids near a naked flame.
11 Keep your apparatus clean and your bench tidy.

Safe laboratory practice
1 **The handling of apparatus**
a *Glassware* – All glass apparatus should be checked for defects before setting it up. Students should always ensure that when they erect an apparatus there is no strain which could damage it. All rubber tubing should be kept free from constrictions, for example, condenser tubing if caught at any point will stop the flow of water and allow inflammable vapour to escape.

If glass tubing is to be inserted into a cork or bung, a hole of the *correct* size should be bored and the tubing, previously rounded off in a flame, inserted with the aid of lubrication. To remove tubing insert a cork borer between the tubing and the bung. If this is not possible cut the bung and remove the tube: *never* force it out. Always use gloves or a soft cloth to protect the hands.

When opening ampoules or bottles with seized stoppers great care is required. Seized stoppers should be gently tapped on both sides with a piece of wood to release them. Failing this a cut is made in the neck of the bottle with a file and a hot wire or glass rod applied to the cut. The latter procedure should be adopted when opening ampoules, and should always be done in a fume cupboard

b *Pipettes* – Safety pipettes of the aspirator type should always be employed when measuring liquids.

c *Desiccators* (*vacuum*) – These are capable of implosion when used under vacuum and, therefore, a safety screen should always be used to protect the operator.

d *Bunsen burners* – These should only be left on when in use. If they have to be left on for some time the air hole should be closed to give the smoky yellow *visible* flame.

e *Electrical apparatus* – Electrical equipment should always be earthed properly and three-core 10 ampere cable used for connections between apparatus and supply. It should never be used with wet hands or when standing on wet surfaces. If liquids are spilled on electrical apparatus it should be thoroughly dried and tested before use. Makeshift connections and slipshod assembly are common causes of accidents and must be avoided.

2 Handling and storage of chemicals
Toxic and explosive compounds will be dealt with in more detail later.

a *Spillages* – Spillages on the bench or floor should be mopped up immediately, after preliminary treatment if necessary (such as soda ash and washing with water for acids, or sand for oily residues). If inflammable solvents are spilled all naked lights should be immediately extinguished. When possible they should be extinguished before using inflammable materials.

b *Carrying of chemicals* – This should be done in a receptacle which will resist breakage or spillage when large bottles or Winchesters are involved or when chemicals are being carried long distances. Otherwise ensure that the student carries bottles with one hand under the bottom and the other round the neck of the bottle.

c *Storage of chemicals*

i Only the minimum amount of chemicals required for immediate use should be kept in the laboratory. The remainder should be kept in a bulk store.

ii Chemicals should be stored according to type and not alphabetically, so that chemicals capable of reacting with one another, for example nitric acid and alcohol, or *sodium*-dried ether and alcohol, are not stored close to one another.

iii Bottles should be carefully and clearly labelled and regularly cleaned. The labels on bottles of hazardous chemicals from reliable suppliers bear warnings of the hazards and these should be transferred along with the name to secondary containers.

iv Chemicals in the laboratory should be made as accessible as possible allowing for safe working. They should not be placed on shelves above eye level and should not be stocked along the centre of the bench unless they are placed on shelves having short projections along the front to prevent bottles falling off accidentally. Wall mounted shelves situated near to the bench, to cut down unnecessary transportation, are more useful. Strong acids and bases should be stored on an inert base such as glass.

v Bulk chemicals should be kept in an outside store and, if possible, inflammable solvents should be kept in an outhouse away from the main building.

vi Chemicals which are subject to decomposition should not be stored in bulk (for example ether, which forms explosive peroxides). Others with high vapour pressures (for example ammonia) should be stored in a cool place away from direct sunlight.

vii Toxic substances, in particular scheduled poisons, should be locked in a cupboard, the key of which is readily accessible to staff but not to students.

3 Protective clothing

The necessity of protective clothing and headgear is not always felt by teachers. Many consider that students are not exposed to risks which are serious enough to warrant the cost of safety clothing. In the short term this assumption may be justified but from the long term point of view this is not so. Students should be encouraged to purchase and wear laboratory coats and a *suitable* supply of safety goggles and gloves should always be available for use when required. For example, if a student is pouring out concentrated sulphuric acid, he should always use goggles and gloves. Many teachers do not allow students to do this but do it themselves or delegate the laboratory technician to do it. In either case the same precautions should be taken. It is imperative that the right kind of example is set by instructors. It is at this stage more than any other that correct laboratory discipline is needed. If a teacher always wears a laboratory coat at this stage students will follow.

4 Laboratory housekeeping

It is essential that students be taught to set up apparatus neatly and to make sure that once set up it is free from hazard. This can only be done if the bench is free of unnecessary impedimenta. All chemicals that are used must be returned immediately to their proper place. It is also essential that students should wash their apparatus immediately after use when they are still aware of the contents. The benches should be cleared and all spillages dealt with, on the floor as well as the bench.

A pair of stepladders should always be available to discourage the practice of using makeshift ladders (such as wooden boxes) to get materials from high places. Two dustbins should be available, one for *glass only* and the other for other laboratory waste.

First aid in the laboratory

First aid in the laboratory should be limited to keeping the patient comfortable until a doctor or ambulance arrives, except in minor accidents.

Minor accidents

Cuts – should be washed thoroughly with water and bound with a bandage or with lint and sticking plaster.

Burns – should be covered with a sterile burns dressing. If blisters form these must not under any circumstances be broken. For mild acid or alkali burns, wash with water, neutralize with dilute sodium bicarbonate or citric acid solution respectively, and cover with a dressing soaked in sodium bicarbonate for acid burns and saline solution for alkaline burns.

Eyes – For low concentrations of irritant vapour in the eyes, wash thoroughly with water. Use water only for washing out the eyes. Mild acid, alkaline, or saline solutions may have precipitated particles which aggravate the irritation. The presence of such solutions can inhibit the more natural, and safer, impulse to rush to a water tap. If in doubt, seek medical aid.

Poison by ingestion – In the majority of cases the poison is spat out by the patient. In these cases, thorough washing with water is usually sufficient. However, common sense must dictate the course of action.

Major accidents

In all major accidents medical help should be sought immediately. All first aid cabinets should contain the names, addresses, and telephone numbers of local doctors, hospitals, and ambulance units.

Cuts – Staunch the bleeding and seek medical aid. If glass or any other foreign body is present in the cut, do not attempt to remove it.

Burns – The initial treatment is the same as for minor burns.

Eyes – Wash thoroughly with water. If the accident is caused by a foreign body in the eye, such as glass, make no attempt to remove it but get the patient to a doctor immediately.

Poison by ingestion – In most cases the cause of poisoning is known to the patient and the first aid treatment referred to in the following pages should be followed. However, if it is not known, ascertain whether there is any burning round the mouth or lips. If no burning is apparent, give the patient an emetic or, as a last resort, place your fingers towards the back of the throat. If burning has occurred give the patient plenty of water or milk.

N.B. Never administer alcohol and never force anything into the mouth of an unconscious person.

Artificial respiration – Although it is extremely unlikely that this will be called for, at least one person and preferably more should be capable of giving mouth-to-mouth respiration.

Emetics – These are given to induce vomiting. The following are the most common types of emetic.
    a   *Mustard.* One teaspoonful of mustard in a glass of warm water. Administer in doses of one quarter of the glass followed by a glass of warm water, repeated at one minute intervals.
    b   *Salt water.* Two teaspoonfuls of salt in water administered at one minute intervals until four glasses have been taken.
    c   *Soapy water.* Shake mild soap in warm water and administer as in (a).

Electrical accidents – When attempting to rescue any person who is electrocuted, turn off the current at the mains or protect the hands by using rubber gloves or some dry material, e.g. duster, jacket. Make sure you are suitably insulated from the floor by standing on a dry mat or a coat. Remove the patient rapidly to a safe place and render first aid.

The following pages contain notes on the effects, and first aid treatment for accidents arising from some chemical substances likely to be encountered in a school laboratory. This is not intended to be an exhaustive list and it should not be assumed that because a particular substance is not included it does not represent a potential hazard. In compiling these notes most of the information given has been obtained from the *Laboratory handbook of toxic agents*, edited by Professor C. H. Gray, obtainable from the Royal Institute of Chemistry. A copy of this book should be available in a known and easily accessible position in every school chemistry department.

Effects

a Extremely irritant vapour
b Extremely irritant liquid (solid)
c Inhalation causes damage
d Ingestion causes damage; represented as follows:
  d, irritation of digestive system and drowsiness
e Highly inflammable

Treatment

A Remove from the immediate vicinity, rest and keep warm
B Wash affected eyes thoroughly with water
C Wash affected parts thoroughly with water
D Ingestion; wash mouth out thoroughly with water. If additional treatment is required it is shown as follows: D, plenty of water to drink, milk of magnesia

Name		
acetaldehyde	a; e; c, headaches and drowsiness; d, drowsiness, damage to digestive system	A; B; D, emetic
acetic acid	c, choking and weeping; b, burns skin, mouth, digestive system	A; B; C, wash with soap and water; D, plenty of water to drink, milk of magnesia
acetyl chloride	as for acetic acid	as for acetic acid
ammonia	c, choking and weeping; b; d, severe internal damage	A; B; C; D, water to drink, vinegar or 1 per cent acetic acid
aluminium bromide (anhydrous)	c, severe irritation and burning of respiratory system. Dust causes severe eye burns. Heat burns on contact with moist skin	A; B; C; D, water to drink, milk of magnesia
aluminium chloride (anhydrous)	as for bromide	as for bromide
aniline	b; d, severe internal damage; c or skin absorption, headaches, drowsiness, cyanosis, mental confusion, even convulsions	A; C, wash with soap and water; D, emetic
barium salts	d, sickness and diarrhoea	A; D, 2 tablespoonfuls of magnesium sulphate, emetic
benzene	DANGEROUS. c or skin absorption, dizziness, excitement, poisonous. Repeated exposure may cause leukemia	A; C, wash with soap and water; D, emetic

Name	Effects	Treatment
benzoyl chloride	a, choking and weeping; d, internal irritation and damage; corrosive liquid burning eyes or skin	A; B; C, wash with soap and water; D, water to drink, milk of magnesia; contaminated clothing washed before re-use
bromine	liquid burns skin and eyes; a; d, local burns, internal irritation and damage	A; D, plenty of water to drink; B; C, wash with dilute ammonia or sodium thiosulphate (not the eyes)
bromoethane	b; d, poisonous; a, choking or weeping; narcotic and anaesthetic effects	A; D, emetic; C, wash with soap and water; contaminated clothing washed before re-use
butylamine	a, choking or weeping; causes skin and eye burns, probably poisonous	A; C; D, water to drink
carbon disulphide	a, eyes; d, poisonous; c, narcotic effects	A, apply artificial respiration if breathing stopped; B; D, emetic
carbon tetrachloride (tetrachloromethane)	DANGEROUS. b; d, sickness and severe internal damage, fatal in small doses; c, headache, depression, sickness, even coma; symptoms delayed for some hours	A, apply artificial respiration if breathing stopped; B; C, wash with soap and water (not eyes); D, emetic; contaminated clothing washed before re-use
chlorine	a, affects eyes and lungs, possible conjunctivitis	A; C
chlorobenzene	c and d, stupor and delayed unconsciousness	A; C; D, emetic
chloroform (trichloromethane)	a, conjunctivitis; d, poisonous; c, giddiness, headache, sickness, unconsciousness	A, apply artificial respiration if breathing stopped; B; D, emetic
chloronitrobenzenes	c or skin absorption, weeping or choking leading to cyanosis and liver injury; d, probably poisonous; skin contact causes dermatitis	A; C, wash with soap and water; D, emetic; contaminated clothing washed before re-use
chlorophenols	c, weeping or choking; d, poisonous; burns skin and eyes	A; D, plenty of water to drink, two tablespoonfuls of Epsom Salts; C, wash affected parts with soap and water; contaminated clothing washed before re-use
copper(II) sulphate	c; d, severe sickness, diarrhoea, intense abdominal pain, collapse	A; D, emetic only if vomiting has not occurred
cresols	c, weeping or choking; d, headache, dizziness, sickness and stomach pain, exhaustion, possible coma; repeated skin contact could cause dermatitis and will burn skin and eyes	A; D, plenty of water to drink, two tablespoonfuls of Epsom Salts in water; B, splashed skin washed with soap and water
1,2-dibromoethane	b; d or skin absorption, poisonous; c, weeping or choking; may have narcotic effects	A; B; C, wash with soap and water, not eyes; D, emetic; contaminated clothing washed before re-use

Name	Effects	Treatment
dichloromethane	b; c, headache, sickness, possible cyanosis and unconsciousness; d, assumed poisonous	A; C; D
ether	c or d, drowsiness, mental confusion, possible unconsciousness	A; B; D, emetic
2,4-dinitroaniline	effects not known; d, assumed poisonous	B; C, wash with soap and water, *not* eyes; D, emetic
2,4-dinitrochlorobenzene	c or skin contact, may cause cyanosis, liver injury; d, assumed poisonous	A; C, wash with soap and water, *not* eyes; D, emetic
2,4-dinitrofluorobenzene	as for above	as for above
2,4-dinitrophenol	c, d, or skin absorption, profuse sweating, fever, shortness of breath, hands and feet coloured yellow; skin contact causes dermatitis	A; B; C, wash with soap and water; D, plenty of water to drink, two tablespoonfuls of Epsom Salts; contaminated clothing washed before re-use
formic acid	a; b; d, internal damage	A; B; C, 1 per cent ammonia in water
hydrochloric acid (hydrogen chloride)	a; b; d, internal damage	A; B; C; D, water to drink; milk of magnesia; contaminated clothing washed before re-use
hydrogen peroxide	irritant, caustic to skin and mucous membrane; d, sudden evolution of oxygen – distension of stomach – nausea, vomiting, internal bleeding	B; C; D, plenty of water to drink; contaminated clothing washed before re-use
hydrogen sulphide	DANGEROUS GAS, irritation of respiratory system and eyes, headache, dizziness, weakness, even unconsciousness	A, artificial respiration if breathing stopped; B
iodine	a; b, solid burns skin; d, severe internal damage	A; B, skin washed with 1 per cent sodium thiosulphate; wash mouth with 1 per cent sodium thiosulphate, drink some, emetic
iron(III) chloride	corrosive in moist state; severe eye burns; thermal and acid burns; c, irritation or burning of mucous membranes; d, severe burns	A; B; C; D, plenty of water to drink, milk of magnesia
methanol	a; b, effect delayed many hours; c, low concentration – headache, nausea, vomiting, irritation of mucous membrane; high concentration – dizziness, stupor, cramps, digestive disturbances; d, damages nervous system (particularly optics system), damages kidney, liver, heart, other organs, can be fatal. Skin contact, dermatitis	A; B; D, emetic; contaminated clothing removed, aired before re-use

Name	Effects	Treatment
2-naphthol	b, skin contact; d or skin absorption, causes kidney damage	A; B; C, wash with soap and water; D, plenty of water to drink, two tablespoonfuls of Epsom Salts
nitric acid	a, b, burns eyes and skin; d, internal damage	A; B; C; D, plenty of water to drink, milk of magnesia; contaminated clothing washed before re-use
nitrobenzene	c, d or skin absorption, difficulty in breathing, vomiting, cyanosis, drowsiness, even unconsciousness	A; B; C, wash with soap and water (*not* eyes); D, emetic; contaminated clothing washed before re-use
nitrophenols	a; b; c, d, or skin absorption, irritation, headache, drowsiness, cyanosis	A; B; C, wash with soap and water (*not* eyes); D, plenty of water to drink, two tablespoonfuls of Epsom Salts; contaminated clothing washed before re-use
orthophosphoric acid	a; b, burns eyes and skin; d, internal damage	B; C; D, plenty of water to drink, milk of magnesia; contaminated clothing washed before re-use
oxalates	a; b; d, internal pain and collapse	B; C; D, emetic
phenol	a; b, skin contact, softening, whitening, development of painful burns, dermatitis with weak solution; skin absorption, rapid, headache, dizziness, rapid and difficult breathing, weakness, collapse; c, long periods – digestive disturbance, nervous disorder, skin eruption, liver and kidney damage; d, severe burns, abdominal pain, nausea, vomiting, internal damage	A; B; C, wash with soap and water; D, plenty of water to drink, two tablespoonfuls of Epsom Salts; contaminated clothing washed before re-use
phosphorus pentoxide	a; b, burns eyes and skin; d, severe internal damage	A; B; C; D, plenty of water to drink, milk of magnesia; contaminated clothing washed before re-use
phosphorus pentachloride	a; b, affects mucous membrane and respiratory system; solid burns eyes; vapour and solid burns eyes and skin; d, severe internal damage; continuous exposure – lung damage	A; B; C; D, plenty of water to drink, milk of magnesia; contaminated clothing washed before re-use
phosphorus trichloride	as for pentachloride	as for pentachloride
silicon tetrachloride	a; b, liquid burns eyes and skin; d, severe internal damage	A; B; C; D, plenty of water to drink, milk of magnesia; contaminated clothing washed before re-use
silver nitrate	b, skin burns; d, internal damage by absorption in blood then deposition of silver in body tissue	B; C; D, emetic
sodium hydroxide	b, severe burns of eyes and skin; d, severe internal damage	B; C; D, plenty of water to drink, vinegar or 1 per cent acetic acid
sodium peroxide	b, thermal and caustic burns on moist skin; d, severe internal damage	B; C; D, plenty of water to drink, vinegar or 1 per cent acetic acid

Name	Effects	Treatment
sulphur chloride	a; b, liquid causes burns; d, severe internal damage	A; B; C; D, plenty of water to drink, milk of magnesia; contaminated clothing washed before re-use
sulphur dioxide	a, causes bronchitis, asphyxia, high concentrations cause conjunctivitis	A; B.
sulphuric acid	concentrated acid, severe skin and eye burns; dilute acid, irritates skin and eyes, dermatitis; d, severe internal damage	B; C; D, plenty of water to drink, milk of magnesia; contaminated clothing washed before re-use. BURNS REQUIRE MEDICAL ATTENTION
thionyl chloride	a; b, causes burns; d, severe internal damage	A; B; C; D, plenty of water to drink, milk of magnesia; contaminated clothing washed before re-use
triethylamine	a; b; d, assumed irritant and poisonous	A; B; C; D, plenty of water to drink; contaminated clothing washed before re-use

Appendix 5
Examinations and tests

The role of examinations in this and any other course should be a secondary one: to measure attainment in the objectives of the course which have *first* been determined on educational grounds. Bearing this in mind, the examiners in the Project tried to ensure that the scheme of examining which they devised could be fitted into the existing examination structure at A-level, and they were constantly aware that what they devised could influence the manner and content of teaching in the future.

There were four main tasks in the field of assessment during the school trials:
1   To provide problems for the students in each topic.
2   To provide examinations for internal use by schools during the trials.
3   To construct a suitable form of A-level written examination in the light of experience gained in the trials.
4   To conduct trials in order to determine a suitable method of assessing attainment in practical work.

The problems for each topic
Many of these problems were used during the trials and amended, where necessary, as a result of comments from the teachers. They are included in the *Students' Book*; the answers, with suggested mark allocations, are given at the end of each topic in the *Teachers' Guide*. There should be sufficient to provide work for the average student, and a wide variety of problems has been included as a guide to teachers who may wish to devise more of their own.

Internal examinations and tests
Five internal examinations were held during the two years of trials. They provided an opportunity to investigate types of question and forms of examination with a view to arriving at the most suitable form of A-level examinations. They also gave teachers a means of assessing the progress of the students, and the Headquarters Team a means of judging the suitability of some of the materials used in the trials. Many of the questions were written by two teams of writers drawn from the two groups of trials schools, and this enabled many teachers to gain experience in writing unfamiliar types of question.

The examinations were assembled by the Headquarters Team, but the answers were marked by the teachers from common mark schemes. The results were analysed by the Research Unit of the London University Schools' Examination Board, whose invaluable assistance is gratefully acknowledged.

The internal examinations did not differ greatly from the form of the A-level examination which is described in the next section. It is not essential that all internal examinations and tests in schools should be in exactly the same form as the A-level examination, but it is desirable that students be given practice in the various types of questions which they will meet in that examination. It is proposed to publish a large number of the questions which were used, to provide teachers with a bank of questions for use in their own tests and examinations.

The A-level examination
The framework in which the examiners had to work consisted of the intellectual objectives of the course, the content of the course, and the various activities the students should undertake during the course. Practical work is an integral part of the course and in many respects it is undesirable to think of it as something separate from the other activities, but it is traditional and convenient to assess attainment in practical work on its own, in addition to some questions on practical work in the written papers.

The intellectual objectives
It has been found that the classification of intellectual abilities which was used in the O-level Nuffield examination can be applied to A-level, although their weighting may well be different:

1 *Knowledge* – the process of remembering.

2 *Comprehension* – the ability to understand and interpret in a routine manner.

3 *Application* – the ability to apply knowledge and understanding to new situations or in an unfamiliar way.

4 *Analysis/evaluation* – the ability to perceive the various parts of a communication, to perceive relationships between the parts, and to evaluate their worth for achieving specified ends.

5 *Synthesis* – the ability to assemble relevant parts to make a complete, complex communication.

The content
The obvious method of classifying the content of the course is by the topics set out in the publications, and this is done. The examiners will expect the candidates to be familiar with the areas of subject matter and principles of these topics. Candidates will be given an opportunity in Paper III of the examination to show that they have studied the wider applications of chemistry outside the basic course, but the examiners will not necessarily expect them to have a detailed knowledge of the supplementary material in the publications.

In many respects, the division of the course into topics is simply a way of dividing chemistry into convenient teaching units and the topic headings do not necessarily indicate the main themes of chemistry. A classification solely by topics runs the risk of neglecting some of these main themes and of overweighting some areas of detailed subject matter. These themes can be classified under seven headings:

1 *Materials exhibit specific behaviour and composition* – This concerns physical and chemical changes in which it is the particular substances which are important. Preparations, analyses, and measurements of quantity, are all included.

2 *Materials exhibit patterns in behaviour and composition* – This will include the Periodic Table, classes of reaction, classes of material, oxidation state, generalizations, and laws.

3 *The explanation of behaviour and composition in terms of structure and bonding* – This includes intra-atomic structure, inter-atomic structure, and the structure of bulk materials.

4 *Kinetics* – how fast changes take place.

5 *Equilibria* – how far changes go towards completion.

6 *Energetics* – the relationship between energy and structure and changes in materials.

7 *The applications of chemistry* – the industrial, medical, economic, and other social aspects of chemistry.

These seven themes are intended to be the essential cohesive elements of the course. They could provide the framework of this and other chemistry courses at this level; and, as they can be illustrated with a variety of subject matter, they could be used as a common basis for the comparison of other courses and examinations. There is some overlap between the classification by themes and by topics, but each classification has a function not achieved by the other and the examiners intend to use both.

The activities
The classification of student activities used at O-level has been adopted for A-level:

A   Consideration of the behaviour of specific substances.
B   Practical work.
C   Using patterns in behaviour and structure.
D   Making measurements and calculations.
E   Using concepts.

Again, there is some overlap with the themes of the course, but classification by activities still serves a useful purpose, if only in ensuring adequate weighting to B and D.

The form of the A-level examination

The examiners have had to compromise between two opposing objectives: one is to measure reliably the performance of all the candidates in the essential objectives and material of the course; the other is to allow candidates to describe and comment on matters in which they are most interested and have studied in greatest depth, whether they are part of the basic course or not. Papers I and II are intended to achieve the first objective, Paper III the second. The following is a brief description of the A-level examination in 1968.

Paper I – This took the form of an objective test of 50 fixed response questions (multiple choice and similar forms) in 75 minutes; there was no choice of question. The types of question were similar to those used at O-level in the Nuffield examination, examples of which are published by the London University Examinations Board in their Pupils' and Teachers' Booklets. More examples will be available when the questions from the internal examinations of the trials schools are published.

Paper II – A common feature of questions in both Papers I and II was that they gave information, sometimes at length, on which the candidates had to work. In Paper II this information was followed by a set of structured questions, each requiring a short answer for which space was left on the question paper. Eight such questions sets had to be answered in two hours. In the 1968 paper some choice was allowed, but this choice is unlikely in the future. The answers were marked from a detailed marking scheme.

Paper III – This paper consisted of a wide choice of questions which candidates could answer freely and at length, but there was one compulsory question on each of the Special Studies which form part of the course. (Candidates are expected to have covered *one* of these Studies.) The answers were marked by impression by two examiners, their marks being added to give the final mark.

The specification and form of the examination was devised by the Chief Examiners in consultation with the Headquarters Team of the Project and a special advisory panel which contained representatives of the Examining Boards. The examination was administered by the London University Schools' Examination Board on behalf of all the Examining Boards.

In the early years of any new examination, particularly when it is examining a new course, flexibility is desirable. For example, a likely extension in the future is a compulsory question on chemistry in action in industry. Such matters as weighting, level of difficulty, timing, and choice of question, must be reviewed from year to year; and it is hoped that a thorough analysis of the results will be available to give guidance to the examiners and their advisers. The Chief Examiners have every hope that their policy of organizing teams of teachers to write the first draft of most of the questions will ensure a continuous flow of new ideas into the examination and will keep the examiners in touch with the development of the course in schools.

Assessment of practical work

The assessment of practical work by teachers has been undergoing trials since September 1966. The purpose of these notes is to explain this form of assessment to teachers who are about to start the Nuffield Advanced Chemistry Course.

The advantages of assessment by teachers

1 The assessment can be done by those who actually observe the students at work over a long period of time, so that practical skills, abilities, and aptitudes, which are not displayed in written work, can be taken into account. In this way the objectives of the assessment can be made wider than those of a practical examination assessed on a single occasion on written evidence alone.

2 Assessment on several occasions is necessary to cover adequately the variety of experiments and techniques which comprise the practical work of the course, and the effect of chance failure or success is less.

3 The assessed practical work need not be limited to those experiments which can be readily administered to a large number of candidates in a limited time. The restriction of practical examinations to such experiments could have an undesirable effect on the variety of practical work in the schools taking the Nuffield course. The form of assessment set out in these notes should lead to a greater freedom of choice of practical work in schools.

4 This scheme gives teachers the opportunity to become involved in the assessment of their students. It could lead to a greater awareness of the objectives of practical work and a situation in which teachers and examiners work together towards these objectives.

The outcome of the school trials

During the two years of trials eight experiments, all of them part of the basic course, were chosen by the Project's Headquarters Team for assessment. Guidance was sent to teachers on the conduct of the experiments, the objectives of each, and the manner in which marks could be awarded. The results for each student were entered on a record card and the record cards were reviewed by the examiners at the end of the first and second years. The main conclusions were as follows:

1 Teachers tended to use a restricted range of marks, and a candidate's total mark was rarely less than the half mark. The mean mark was correspondingly high. From an analysis of the marks, coupled with the direct observation of practical classes by the Headquarter's Team, it can be inferred that the spread of ability is less in practical work than it is in written work, and it is unlikely that a student can be said to have failed in his two years of practical work.

2   The objectives of practical work were the subject of much debate. Finally, all the teachers who took part in the trials were invited to state what they considered the objectives to be. The objectives and their weighting, which are set out later in these notes, follow closely the outcome of this inquiry.

3   Many teachers felt that the detailed instructions, given by the Headquarters Team for each of the assessed experiments, unduly restricted them in their assessment. It became clear that no single set of instructions would meet the wishes of all the teachers, and it was decided that the proper course was to give as much freedom as possible to the teachers in the choice of experiments for assessment and in the conduct of the assessment. Some uniformity is necessary, however, in the objectives, the areas of work, the type of experiment, and the form in which the results are recorded.

4   It has been a constant policy of the trials that the assessment should be conducted on normal school work and be as unobtrusive as possible, but often there was a noticeable difference in the attitude of a class to the work when they knew it was to be assessed. Some teachers welcomed the more diligent work; others regretted the more formal atmosphere.

5   Teachers were rightly concerned lest the raw mark which they awarded their students would be the one used in the final A-level mark. The raw marks were used by the examiners to determine only the distribution of marks within a teaching set, and this policy will continue.

6   Possibly the most difficult decision which had to be made was whether there was sufficient difference in the general standard of practical work between teaching sets and between schools to call for the use of a moderating instrument in order to make adjustments to the marks. For the time being, it seems that some form of moderation is desirable, but the matter will be reviewed from year to year.

7   Teachers found that those objectives which could be marked from written evidence were easier to assess than those which called for direct observation by the teacher during an experiment. It was found to be very difficult to assess more than one of the latter during any one experiment. It is hoped that assessment by direct observation will become easier as teachers become more familiar with it and can choose those experiments in which they know they can pick out particular faults and virtues without interfering with the normal teaching work.

The form of assessment after 1968

The objectives
The following objectives, with weighting given as a percentage, are suggested.
1 Skill in observation (25 per cent)
2 Ability to interpret observations (15 per cent)
3 Ability to plan experiments (10 per cent)
4 Manipulative skills (30 per cent)
5 Attitude to practical work (20 per cent)

In many respects the form of assessment advocated in these notes has still to be evaluated, and it would be wrong to regard this statement of objectives and weighting as fixed for all time. Teachers may wish to modify both the objectives and the weighting, particularly in the finer divisions of the objectives. But for the time being it will help in the evaluation of this form of assessment if teachers can keep as close to the suggested pattern as possible. If teachers feel that they must deviate from this pattern they will still be asked to show on the record cards how they allocated marks to particular objectives.

The objectives, 1 to 5, provide the broad basis for the final assessment, but for assessment on particular occasions a finer division is necessary. The following notes on the objectives may be helpful, but teachers should not feel that they are rigidly bound by them.
1 An assessment of the ability to observe leans heavily on a student's ability to report, and it is inevitable that most of it will take the form of awarding marks for written records of observation. But there are two aspects of observation for which teachers can give credit while a conventional practical examination cannot, and it is hoped that neither will be overlooked. The first is oral reporting which is usually more immediate and less rehearsed than written reporting. The second is a student's observation and interest in unusual and unexpected features of an experiment even though they are not relevant to the immediate purpose of the experiment. It will not be easy to put a mark on these two aspects but at least credit could be given in a periodic review and grading of each student's attitude to practical work.
2 The assessment of *interpretation* is similar to that of observation in that teachers can only judge what students write or say. It will give credit for an analytical approach to observations, including unexpected observations and the ability to comment critically on practical situations.
3 The ability to *plan* is most easily assessed by asking the students to write their plan and submit it for marking before they start, but this can only be done in those experiments for which information to students on procedure is restricted. It should include the ability to recognize sources of error, and the ability to make predictions about other experiments.

It can be argued that both interpretation and planning can be assessed on written examination papers. This may be so, but a student's interpretation of observations which he has actually made, and his plans for an experiment which he knows he must try to put into operation, may well be different from his interpretation and planning of unreal situations on paper. The thinking processes may be similar but their context is different, and therefore their outcome may be different.

4   *Manipulative skills* can be assessed by marking the results of experiments, for example yield, purity, and accuracy (accuracy of work rather than accuracy of calculation); they can also be assessed by direct observation of such things as orderliness, methodical working, dexterity, and speed. It is desirable that this assessment should not be confined to routine operations; it should include the ability of students to adapt their manipulative skills to new situations and to follow unfamiliar instructions.

It may be possible for teachers to use specific experiments for assessing some parts of manipulative skill, particularly yield, purity, and accuracy. However, many teachers feel that a grade based on a general impression of work over a period of, say, a term or even a year is more appropriate for the less exact aspects of manipulative skill which can only be assessed by direct observation. If periodic grading is used, it is important that teachers should consciously look for evidence of the skills throughout the period and not simply rely on their memory at the end of the period. (A further note on grading by impression is given later.)

5   For the purpose of this assessment *attitudes* can be divided into three categories:

a   Willingness to cooperate in the normal routine of a chemistry laboratory. This will include a knowledge of safety regulations and a willingness to comply with them and the other regulations which are necessary for the efficient running of a laboratory.

b   Persistence and resourcefulness – the will to work unaided on set problems and to see them through to a meaningful conclusion.

c   Enthusiasm; a commitment to practical work as a worthwhile pursuit without compulsion. This could show itself in suggestions for lines of investigation not specified in the course and in assuming the more active role when working in groups.

A grading at the end of each year, or even at the end of the course, is probably the most appropriate method of assessing attitudes.

Areas of subject matter

The assessment in the future will not be res ricted to specified experiments as it was in the trials. Teachers will be free to choose which experiments to use and these experiments need not form part of the published Nuffield course. But in order to get a reliable order of merit it is important that the main areas of subject matter are covered. On each student's record card the examiners will expect to see a record of assessment on at least *eight* experiments, including at least *one* from each of the following areas of subject matter:

1 Changes in substances and patterns in changes in substances. (This is a deliberately broad category to include most of the course not covered by 2, 3, and 4, below. Most of the assessed experiments are likely to come in this category.)

2 Equilibria.

3 Kinetics.

4 Energetics.

Types of work
Three main types of work can be identified in the course and the examiners will expect to see each one used in the assessment, but not necessarily with equal weighting:

1 Quantitative work.

2 Preparative work.

3 Qualitative work.

Allotment of marks
The number of occasions on which marks are allotted for practical work is left to the discretion of the teacher provided that the minimum number of assessed experiments is attained. It is desirable that the objectives, with the possible exception of objective 5, are assessed on more than one occasion in order to get a reliable estimate of ability.

Students should be told at the beginning of the course that their practical work will be assessed and that the results will be sent to the examiners. They should also know, in general terms, the qualities for which the teacher will look. They should be told that the assessment will be more obvious on some occasions than on others, but that it will usually take place during normal work and not on special occasions. It may reassure them to know that the average mark for practical work will be higher than that for the written examination.

The examiners will not necessarily expect that all the students in a teaching set will be assessed on the same experiments, but an equal balance of objectives is desirable. Assessment of objectives 1 and 2 will be difficult in experiments in which the students work in groups, but teachers should not feel restricted in the assessment to those experiments which are performed by students working alone: some assessment of objectives 3, 4, and 5, is possible in group work.

During the trials a grading scheme (5–4–3–2–1) was used to allocate marks to a particular objective on a particular occasion, and teachers may find a five point scale useful especially for those parts of the work which are marked by impression rather than by a mark scheme, but they are free to use other schemes if they wish.

There are three main ways in which marks can be allocated and teachers will probably find that they have to use all of them:

1 *A mark scheme* – This will be most useful when marking written evidence of observation, interpretation, planning and accuracy.

2 *Marking by impression on a single occasion* – This will be useful for marking evidence which is less precise than that mentioned in 1. For example, a teacher may wish to assess dexterity in handling unfamiliar apparatus – say in chromatography. This can best be done by direct observation of the students at work, and in a large class it is not easy. Teachers should try to assess only one such quality by impression during one experiment. Furthermore, teachers are usually too busy to do more than make a note of those who are doing well or badly; the rest can only be given the mark for average work.

A good deal of thought during the trials was given to the question of whether 'average' should be the teacher's estimate of the average for his set or his estimate of the average for A-level work in general. The former has the disadvantage that a teacher cannot estimate the general level of practical ability of his set until he has been teaching them for some time. The disadvantage of the latter is that it depends on the teacher's experience of A-level work. Since the raw marks will be scaled to a norm anyway, it is their distribution and the internal consistency of the marking which are important. But some uniformity is desirable and it has been decided that teachers should try to judge what is average from A-level work in general rather than from the work of their teaching set.

Thus:

excellence of work unusual in an A-level student	5
above average work	4
average work	3
below average work	2
unusually poor work for an A-level student	1

3 *Marking by impression after a period of time* – This will apply mainly to attitudes to practical work but teachers may wish to assess other objectives, particularly some of the less precise aspects of manipulative skill, periodically rather than in single experiments. The period is left to the teacher's discretion; it could be done once a term, once a year, or, for some objectives, once at the end of the course. The criteria for awarding a grade are those set out in (2).

Information required by the examiners

Three or four weeks before the written examination the examiners will require the following information from schools.

1 All the record cards of the students arranged according to teaching sets. A sample showing how a record card can be completed is given in Example A; this should not be taken to be an example of an ideal assessment.

2 A summary of the marks for practical work for each set, showing the total mark awarded to each candidate as a percentage (Example B).

3 Any other relevant information, for example:

a Outstandingly good or bad performance throughout the course by individual candidates.

b Outstandingly good or bad performance by a set as a whole.

c Circumstances such as absence or change of school which could affect the performance of a candidate, particularly if it has resulted in the completion of fewer than eight experiments in which assessment has taken place.

The moderation of marks

The present need for moderation of the teachers' marks has already been mentioned. For various reasons, including those of ineffectiveness and expense, moderation by visiting examiners, moderation by a practical examination, and moderation by performance on the written papers as a whole, have been rejected.

Moderation will take place by the performance of teaching sets as a whole in those questions in the written Papers I and II which will have been designed to test practical work; these questions will form not less than 20 per cent of the two Papers. It seems that performance in these questions is more likely to reflect the general level of practical work of a teaching set than any other moderating instrument which can be conveniently applied.

In 1968 the form of moderation was as follows. Three ranges of marks were allocated, based on the performance of teaching sets as a whole in the written practical questions. To conform with the findings of the trials, all three ranges of marks were narrow, and overlapped, and all three mean marks were higher than the half mark. The maximum mark for the teacher assessment of practical work was 70, which comprised 14 per cent of the maximum mark for the whole examination and the three mark ranges were:

top range (20 per cent of teaching sets), mean mark 57.5, standard deviation 4
middle range (60 per cent of teaching sets), mean mark 47.5, standard deviation 4
bottom range (20 per cent of teaching sets), mean mark 37.5, standard deviation 4

The form of moderation will be reviewed from year to year.

In conclusion

This is a first venture in this form of assessment of A-level practical work in chemistry by teachers, and in many ways it is still on trial. At first it will no doubt seem strange to both teachers and students, but it is hoped that in time it will become accepted as part of normal sixth form work. The examiners will welcome queries, comments, and suggestions from the teachers who operate the scheme.

Example A
Record cards showing imaginary marks for a student

N.B.   The reference numbers and titles of experiments are based on the written materials before publication.

Name.. School.................... Form and set..................

NUFFIELD FOUNDATION SCIENCE TEACHING PROJECT
CHEMISTRY ADVANCED LEVEL
ASSESSMENT OF PRACTICAL WORK

Duration of course: September 19.... to July 19....

Record Card Number $\frac{1}{.....}$

Assessed experiments and other assessments		Objectives (showing number) and marks (showing maximum)	
2.3a	Properties of chlorides	Observation (1) 6/10	Interpretation (2) 2/5
5.3c	Iodine/Iodate reaction by titration	Accuracy (4) 7/10	
6.1b	Investigation of some compounds of s-block elements	Observation (1) 8/10	Interpretation (2) 3/5
7.2b	Enthalpy of hydration	Comment on sources of error (3) 3/5	
9.4b	Preparation of phenyl benzoate	Yield (4) 2/5	Purity (4) 4/5
—	General impression of manipulative skill (year end)	Orderliness and methodical approach (4) 4/5	Speed (4) 3/5
—	General impression of attitude to practical work (year end)	Routine (5) 4/5	Persistence (5) 4/5

Summary

		Objectives					
		1	**2**	**3**	**4**	**5**	**Total**
....	Marks						
Signature of teacher	Scaled marks						

Name....................................... School .... Form and set ....

NUFFIELD FOUNDATION SCIENCE TEACHING PROJECT
CHEMISTRY ADVANCED LEVEL
ASSESSMENT OF PRACTICAL WORK

Duration of Course: September 19.... to July 19....

Record Card Number: 2
....

Assessed experiments and other assessments		Objectives (showing number) and marks (showing maximum)		
10.2a	Determination of strength of hydrogen bond	Planning (3) 5/10	Interpretation (2) 2/5	
12.3	Solubility product of silver iodate	Accuracy (4) 4/5		
17.3a	Prediction of redox reactions	Prediction (3) 3/5	Observation (1) 6/10	
14.2	Rate of reaction: acetone and iodine	Recognition of errors (3) 4/5		
16.2a	Investigation of some complexes	Observation (1) 4/5	Interpretation (2) 3/5	
18.3a	Chromatographic separation of amino acids	Dexterity (4) 3/5	Observation (1) 4/5	Interpretation (2) 2/5
—	General impression of manipulative skill (year end)	Orderliness (4) 4/5	Speed (4) 4/5	
—	General impression of attitude to practical work (year end)	Routine (5) 4/5	Enthusiasm (5) 4/5	

Summary

	Objectives					
	1	**2**	**3**	**4**	**5**	**Total**
Marks	28/40	12/25	15/25	35/50	16/20	
Scaled marks	18/25	7/15	6/10	21/30	16/20	68/100

....
Signature of teacher

Example B
An example of the form of the summary of assessment for a teaching set

School ....   Form and set ....

NUFFIELD FOUNDATION SCIENCE TEACHING PROJECT
CHEMISTRY ADVANCED LEVEL
SUMMARY OF ASSESSMENT OF PRACTICAL WORK

Duration of Course: September 19.... to July 19....

Candidate number	Name and initials	Total mark /100	Leave this space blank

Signature of teacher... Date..

Appendix 6

Programmed learning

The need for the highest quality and quantity of educational opportunities has in recent years led to a growing interest in new teaching and learning methods. This has come about partly as a result of the need to teach more students, and partly because of increasing dissatisfaction with the content and method of much of the teaching that has gone on in the post-war period. Not only what is to be taught, but how it can best be taught, how the results of teaching can be made evident, and how the individual student can best be brought into the centre of the process, are the problems which the new methods have made it their aim to solve. What has emerged is a firm emphasis on clarification of learning objectives, reliable assessment of results achieved, participation of the student in the process of learning, and the greatest degree of individual instruction.

Research into programmed learning with its basis in behavioural psychology has made significant contributions towards the practical implementation of these aims, and has led attention back to the best of traditional methods. We now recognize that there is relatively little difference between a good programme and the best teachers in the way that they both confront students with new knowledge and attitudes. However, pressure of numbers, and the avalanche of information which falls on us every day, mean that it has become more difficult for even the best teacher to involve his students in the newest information and yet give each of them the individual attention which they rightly demand. Systems of programmed instruction help the teacher to overcome this problem and channel his own energy more efficiently.

The personality of the teacher, and his ability to adjust to the immediate needs of his students, are among the obvious advantages of good classroom teaching, but there is a limit to what the teacher can do personally with classes of more than three or four students. A learning programme, on the other hand, generally allows a student to work at his own pace, and it frequently contains special sections, or 'branches', which can be used as a further aid by the individual student. Moreover, a programme is tried out with suitable groups of students and revised until satisfactory results are obtained. This 'validation' procedure which means that achievement is written into the programme, and the flexibility to adjust to the needs of individual students, are important features of programmed learning not present to the same extent in any other form of teaching.

The best programmes and the best orthodox teaching have much in common, and can usefully complement each other. Indeed, research has shown that the most effective way of teaching and learning comes from a combination of both methods. In essence, programmed learning provides a tool which teachers can use to increase the overall effectiveness with which they instruct and their students learn. Practical experience during programme evaluation in the Nuffield Advanced Chemistry Trials' Schools showed that teachers generally used the programmed texts in just this way; they did not regard them as competitive elements.

Successful integration of this sort needs careful planning and the right kind of programme. The relatively slow progress of programmed learning in schools and universities has been partly due to lack of money and the concentration of some of these scarce resources on the development of teaching machines rather than a wide range of programmes. The chief difficulty has been the lack of good programmed material which could be used with courses in a number of different ways and with students of a fairly wide range of age and ability. These are essential requirements if a programme is to be widely used as a teaching and learning aid. In this section of the Nuffield Science Teaching Project we have investigated the extent to which this can be done with courses based on a new approach to what is taught. The development and structure of these programmes, and their use by schools engaged on trials of the main chemistry materials, are discussed in this appendix. A few explanatory notes on programmed learning are also provided.

Programmed learning and teaching machines

A properly constructed learning programme of proved effectiveness is the essential basis of this kind of teaching. Programmes can be presented in a number of ways, for example, as a book, or a film in a teaching machine, or on tape. Most programmes can be used for self-instruction, but increased effectiveness is achieved if the teacher involves himself with the work.

The objectives and content of any part of the course must be analysed in detail before the programme is devised. A test called a *'post-test'* or *'criterion test'*, which checks whether the instructional objectives have been attained, is an essential part of a programme. The material is broken down into steps (*frames*) and arranged in a suitable learning sequence. Some programmes are *linear* (all learners work through the same sequence of frames). In many cases, programmes include at least some *branched* sections for remedial or other purposes. For example, students may be asked to choose between a number of alternative answers; their responses then determine in which sequence the rest of the programme will be presented to them.

The learner usually works through the programme individually, and at his own pace, although working in pairs or larger groups can work well in certain circumstances. The student is required to make a *response* at each frame – answer a question, perform some operation, draw a conclusion, choose between a number of alternatives, and so on. He is then usually informed whether his response was the correct one.

A programme should not be offered for general use before it has been tried with a representative sample of learners and revised until a large proportion of the sample are able to achieve an acceptable score.

In this way, a properly validated programme offers a reasonable assurance of successful learning, but a programme should also be assessed in a wider sense (*evaluation*) to decide whether its objectives are worthwhile, appropriate in a given situation, and reached in an efficient manner.

The time that an individual student will take varies, but it should on average be reasonable when compared with how much has been learnt. A direct comparison between the average time taken to work through a programme and the time which it would take to get the same material across in the ordinary way can be misleading, since conventional lessons generally require further activities, like homework, before the material is learnt.

At first, programmed learning was equated with the use of a *teaching machine*. This gave rise to much hostility among teachers and instructors. In fact, a machine provides only one way of presenting a programme, while a book is another.

Experiments generally indicate that machines are not superior to programmed texts. This is not surprising since in most cases, machines have been merely used as 'page-turning' devices, which may in addition prevent the learner from 'cheating' by reading the answer first. In general students do not cheat once they realize that they are entirely responsible for their own progress while working through a programme.

A teaching machine, however simple, may have advantages in certain situations, for example in a laboratory or workshop to keep the programme clean, or with low ability students who may have developed a hostility to books. A book is generally cheaper and can be used for working at home. On the whole, machines present real advantages only when their mechanical or electronic potential is fully utilized.

Programme writers are beginning to make more and more use of the full range of audio-visual aids and computer facilities to present their programmes. It is up to teachers and educationists to make sure that the use of the so-called 'multi-media' approach does not get out of hand and lead to the inefficient or inappropriate use of machines and media.

Programmed learning has made important contributions to recent educational developments. Some of the original 'principles of programmed learning' based on the pioneering work of Skinner and Crowder are being re-examined, but the real core of programmed learning – namely clearly stated objectives, the critical evaluation of achievement, and a way of using information and results to amend and improve a system geared as far as possible to individual needs – is surely common to all good teaching. Programmed learning is one of the most effective means by which these methods can be made available to the maximum number of teachers and students.

Nuffield Advanced Chemistry Programmed Texts
The initial *objectives* of the work on programmed learning were:

1   To produce a number of programmed texts which could form an integral part of the course. (No attempt was made to programme the whole course.)

2   To include at least one programme which uses the Nuffield approach to experimental work by investigation.

3   To show that these programmes 'filled a need', and would be welcomed by teachers within the Nuffield Trials' scheme, achieving acceptance by those who used them, and good results from students.

Conclusions based on this study include:

1   A high proportion of Nuffield Chemistry teachers used the programmes willingly and were able to integrate them into the course.

2   Results (test data) and comments from teachers and individual students showed that the programmes had worked well.

3   Programmes need not be confined to factual material or a didactic approach.

4   A good deal of flexibility can be incorporated into programmed texts. They can be used to replace normal classwork, for revision, to help students who have fallen behind and to extend the knowledge and background of others, with students of varying ability and working at different rates, in class or at home, for theoretical as well as practical work.

5   Teachers usually encouraged students to use the programmes out of normal classtime. Programmed texts can be used at home, and have a positive advantage over teaching machines in this respect.

6   Regular class discussions, and similar activities in which teachers can become involved with their students' work, are of considerable value. Student motivation and interest is enhanced if programmes include up-to-date scientific and technological information as well as optional sections including more advanced work.

7   The structure of a programme must allow teachers and, as far as possible, students to use it in whatever way they consider relevant. The programme must essentially be a system of learning which can be adjusted to different needs.

Development and structure of the programmes

Programmes were sent out to trial schools only 'on request'. Teachers were each sent a sample copy and asked to request students' copies if they wished to use them. It was hoped to obtain in this way some measure of the extent to which teachers considered the programmes a useful part of the course.

The programmes were in fact used by students in all but one of the Advanced Chemistry Trials' Schools. They were also tried out in a small number of other schools, training and technical colleges.

The results of tests, and the comments from individual teachers and students, helped to eliminate specific difficulties, and, more important, to develop an overall structure which with even greater flexibility could gain wider acceptance.

As a result of this work, four programmed texts are available:
Amount of substance   The mole concept and its use in solving problems. For use with Topic 1 (and to some extent Topics 2–4).
Oxidation numbers   An introduction to oxidation and reduction. For Topic 5.
Names and formulae of carbon compounds. Topic 9.
Ethanol and other alcohols   An introduction to the reactions of carbon compounds. Topic 9. This programme adopts a fully investigational approach integrating practical and theoretical work – see objective 2 on page 330.

Each programme contains:
1   *A number of separate programmed sections,* which include optional 'branches'. Some of these branches are intended as remedial work, others as extension work to widen students' general background knowledge and understanding.
2   *Summaries* for each section with references to programmed sections.
3   *Review problems* for each section (essentially 'post-tests'). Optional problems are also included.
4   *Answers to review problems* with explanatory notes and a marking scheme.

5   *Remedial programmed exercises* to be worked through by students who fail to reach a specified standard in their answers to review problems.

6   *Extension material* (some not in programmed form). This extends the general scope of the texts and varies to some extent in each case.

For convenience, each programme has been divided into two parts.

The programme for Topic 9, which is mainly based on practical work, has a slightly different format. In the first version of this programme, answers to questions resulting from the experimental work were given immediately after each question. Students found it difficult not to look at the answers before performing the experiments. These answers were therefore separated entirely from the questions and this was found to be satisfactory. Equations relating to the experiments are separated from the main programmed sections. This allows teachers to encourage students to speculate about the precise nature of the reactions involved. An introduction to simple reaction mechanism is given in some of these separate sections. Optional experiments are also included. A list of chemicals and apparatus needed for the experimental work is included in the *Apparatus and materials list* obtainable from the A.S.E., College Lane, Hatfield, Herts.

It must be stressed that programme writing involves a searching analysis of subject matter, precise planning of detailed teaching strategy and critical assessment of achievement. Examination of the actual programmes will show that they are not identical in content to the corresponding sections in the *Students' Book*. Practical use in schools strongly indicated that the programmes offer an acceptable alternative to the *Students' Book* material, that they have positive advantages in some situations, and can be used as additional aids for students in other cases.

Effective use of the Advanced Chemistry Programmed Texts

Organizing the use of programmes so that they fit into a school course and reinforce other forms of instruction is as important as the actual structure and contents of the programmes. The Advanced Chemistry Programmed Texts were developed in close collaboration with the Trials' Schools, and successful integration into school courses was the main aim of the work.

Some important points arising from experience with programmes in Trials' Schools are summarized below as answers to specific questions.

1   *Can the programmes be used by the students alone without any help from the teacher?* Yes, but students should be encouraged to ask for assistance, if

required. The involvement of the teacher in the work is an additional encourage-
ment for the student.

2 *Is the content of the programmes identical to that of the corresponding
sections in the Students' Book?* The reasons why this was not done are indicated
in the last paragraph of the previous section on 'Development and structure of
the programmes'.

3 *Can the programmes be used to introduce a topic, as well as to revise and
extend classwork?* Both methods are applicable. Teachers vary in their approach:
(*a*) 'The programme served as a good introduction, since it was simple enough
to follow without any help from me. This introduction helped the pupils so
much that they found succeeding classwork rather easy.' (*b*) 'I want to introduce
the topic myself leaving the programme to cope with revision and pupils who
had been absent for lessons.' In early programme trials, both methods were
used about equally, later a preference for using programmes to 'introduce'
rather than to 'revise' emerged. Students, too, seemed to prefer this method.

4 *Are the programmes more effective than ordinary classroom instruction?*
In some instances teachers felt the programmes were more effective (for example
'Quite frankly, I think the programme taught this topic more efficiently than I
could have done, and I shall use it again.') This is difficult to prove by controlled
research. Essentially teachers use programmes as worthwhile aids in given situations (see 5 below). Active involvement of the learner while using a programme
is an acknowledged advantage, and many students commented on this spontaneously.

5 *How were the programmes used in the Chemistry Trials' Schools?* Most
generally the programmes were set as homework, thus saving valuable time in
class. Examples of actual uses of the programmes are: (*a*) normal classwork
done at home; (*b*) programme used as preparation for later classwork; (*c*) theoretical programme worked on in class while ordinary practical work was in
progress; (*d*) programmed 'test-tube experiments' performed while working on
organic preparations; (*e*) programmes used by fifth formers as preparation for
sixth form work; (*f*) programmes used by upper sixth as revision of first year
work. Programmes were also used, for example, in the absence of teachers or
of students, after transfer between schools, working after normal school hours,
and during the last week of term when no organized work could be done, as
additional work during vacations, and in other situations where conventional
teaching is difficult.

6 *What is the best way of using the programmes?* This varies, but points to
bear in mind are: (*a*) Students who have had no previous experience of programmed learning will benefit from an introductory talk, which stresses that
they will be responsible for their own learning. (*b*) It may be best to work
through the first few pages with the students to sort out any difficulties and
queries. (*c*) Most students find it difficult to concentrate after 30–45 minutes of

non-experimental programmed work. Suitable points for interrupting work are clearly indicated in the programmes. (*d*) Progress may be checked by working through review problems in class, or by discussion. (Further hints are given in 7 and 8 below.)

7 *Is it necessary for any set of students to work through the programme as a whole?* No, the programmes have been deliberately divided up into short and reasonably self-contained sections.

8 *Is it best to spread out the use of a given programme over a period of some weeks?* Generally, yes. One teacher, for instance, commented: 'The programme is excellent background work, a sort of ' teach-yourself topic'', done concurrently with other work. Another year, I feel, that once my set has finished the "Mole" programme, I would start giving them the "Oxidation Numbers" programme and suggest they try and finish it before we start Topic 5. Then the discussion of the ideas in this topic would have real value.'

9 *Should teachers mark answers to* (a) *frames,* (b) *review problems?* Students are not required to write down answers to frames in every case and can generally be left to cope on their own. It is, however, desirable to check on progress in some way, for instance, by checking answers to review problems.

10 *Should students be supervised while doing programmed practical work?* An obvious precaution. No special safety problems were encountered during trials of the Topic 9 practical programme.

11 *How can the programmes overcome difficulties, such as differences in ability, interest, background knowledge, and rate of progress of students?* Straightforward 'linear' programmes generally cannot cope with these points. The Advanced Chemistry Programmed Texts were developed to provide practical solutions to these problems. For instance:

a *Ability differences* – Able students can sometimes work from summaries only, without using the fully programmed sections. Optional problems and extension branches are provided. Less able students can work through the remedial branches and programmed exercises.

b *Background knowledge* – A special appendix is provided for instance in one programme.

c *Rate of progress* – This is not a problem if programmes are used outside normal classes. Optional problems and experiments can be worked through by the faster students in class.

d *Interest* – Optional enrichment information is given.

12 *How can a teacher make a rapid initial assessment of the potential value of one of the programmed texts?* Read the summaries first (subject content), then review problems. Next, part of each section might be read to assess teaching method, general approach, and so on. Finally, examine the additional material given at the end of each text. (Time permitting, work through the whole programme.)

13 *Do programmes tend to encourage a limited approach to a topic, by pre-senting a kind of 'capsule' of condensed learning?* A deliberate effort to avoid this has been made in the chemistry programmed texts. Information of general scientific and technological interest, occasional references to original work, 'open-ended' extension material has been included. Students commented fa-vourably on this, e.g. 'The programme related the work to real life situations, essential if the work is to be taken with a lively interest. This also removes the "stuffiness" from learning, one is not just learning for exams.' Some students stated that they were stimulated to further reading.

14 *What is the attitude of students to programmes?* Much depends on how successfully the programmes were combined with other work. One student for instance commented: 'This programme provided a firm foundation for later lessons in the Nuffield scheme and did not detract from their interest.' Some students found programmes efficient but 'a little slow and monotonous', others 'enjoyed it all and did the optional problems for pleasure'. The personality of the student is probably important. Some students considered 'programmes helpful, because they give one confidence'; others seemed to like the most diffi-cult sections best. Students were not asked specifically whether they preferred to be taught by a programme or by a teacher, but a few provided spontaneous comments (approximately half of these preferred the programme).

15 *Are any other published programmes likely to prove useful for the course?* Early programmes tended to concentrate on learning by rote rather than under-standing, and to be out-of-date in their scientific content and teaching approach. This situation is changing rapidly. A full list of programmes is published by the Association for Programmed Learning in their *Yearbook of educational and instructional technology, 1969/70, incorporating Programmes in Print* (available free to members of the Association). Useful articles have appeared in journals such as *Visual Education, Education in Chemistry,* the *Journal of Chemical Education,* and the *Times Educational Supplement.*

Teachers should scan programmes possibly in something like the way suggested under 12. Programmes which do not have a 'post-test' are on the whole less useful, as it is difficult to assess what the programme is supposed to achieve. In general, the accuracy and relevance of what is taught, the teaching method used, evidence from trials with representative groups of learners, and the presence of a post-test are important points to check.

16 *Is it possible for teachers to write their own programmes?* Programme writing is a skill which even with experience takes time and effort. On the other hand, it is very like the careful preparation of lessons. Teachers, who have had some experience of using programmes, may well wish to write short programmes for use in their own schools. Local groups of teachers often co-operate in this effort. Courses in programmed learning are available at many centres.

It is generally agreed that programme writing is a useful educational discipline. It fosters a critical attitude to the immediate objectives of the teacher, as well as to the wider issues of what a course and curriculum should contain.

Reference sources and bibliography

Throughout the advanced chemistry course, reference is constantly made to the Nuffield O-level Chemistry Project publications, and no further recommendation should be needed here. Of these O-level materials, however, one which is specially recommended as a general reference source is the *Handbook for Teachers*. This not only contains a wealth of useful information and ideas, but it also has at the ends of many chapters extensive lists of references and suggestions for further reading.

The other works of reference and useful books which are listed below are intended as a general guide to some sources of further information which teachers may from time to time need to consult. Specific references to individual papers are not included, since these are given at the appropriate point in the text. There are also some books to which the teacher or student is referred for further reading on particular aspects of individual topics: these are listed both here and in the topics concerned.

The works listed here are arranged under a series of subject headings which are arranged alphabetically. This list is not claimed to be exhaustive: it is merely a collection of some sources which were found helpful.

Applications of chemistry
Nuffield Advanced Chemistry (1971) *The chemist in action.*
Raitt, J. G. (1966) *Modern chemistry: applied and social aspects*, Edward Arnold.
Ives, D. J. G. (1964) *Principles of the extraction of metals*, Monographs for teachers No. 3, Royal Institute of Chemistry.
Chilton, J. P. (1968) *Principles of metallic corrosion*, Monographs for teachers No. 4, 2nd edition, Royal Institute of Chemistry.
Samuel, D. M. (1966) *Industrial chemistry – inorganic*, Monographs for teachers No. 10, Royal Institute of Chemistry.
Samuel, D. M. (1966) *Industrial chemistry – organic*, Monographs for teachers No. 11, Royal Institute of Chemistry.
Kilner, E. and Samuel, D. M. (1960) *Applied organic chemistry*, Macdonald and Evans.

Chemical energetics and equilibrium

Millen, D. J., ed. (1969) *Chemical energetics and the curriculum*, Collins.

Bent, H. A. (1965) *The second law*, Oxford University Press, New York.

Nuffield Advanced Physics, Teachers' Guide to 'Change and Chance' (in preparation).

Ashmore, P. G. (1961) *Principles of chemical equilibrium*, Monographs for teachers No. 5, Royal Institute of Chemistry.

Guggenheim, E. A. (1966) *Elements of chemical thermodynamics*, Monographs for teachers No. 12, Royal Institute of Chemistry.

Caldin, E. F. (1958) *An introduction to chemical thermodynamics*, Oxford University Press.

Denbigh, K. (1966) *The principles of chemical equilibrium*, 2nd edition, Cambridge University Press.

Rushbrooke, G. S. (1949) *Introduction to statistical mechanics*, Oxford University Press.

Kauzmann, W. (1967) *Thermodynamics and statistics*, Benjamin.

Harvey, K. B. and Porter, G. B. (1963) *Introduction to physical inorganic chemistry*, Addison-Wesley.

Allen, J. A. (1965) *Energy changes in chemistry*, Blackie.

Campbell, J. A. (1965) *Why do chemical reactions occur?* Prentice-Hall.

Firth, D. C. (1969) *Elementary chemical thermodynamics*, Oxford University Press.

Chemical kinetics

Ashmore, P. G. (1967) *Principles of reaction kinetics*, Monographs for teachers No. 9, 2nd edition, Royal Institute of Chemistry.

Bond, G. C. (1963) *Principles of catalysis*, Monographs for teachers No. 7, Royal Institute of Chemistry.

Campbell, J. A. (1965) *Why do chemical reactions occur?* Prentice-Hall.

King, E. L. (1964) *How chemical reactions occur*, Benjamin.

Laidler, K. J. (1963) *Reaction kinetics* (2 volumes), Pergamon.

Latham, J. L. (1962) *Elementary reaction kinetics*, Butterworths.

Chemical nomenclature and units

McGlashan, M. L. (1968) *Physico-chemical quantities and units, The grammar and spelling of physical chemistry*, Monographs for teachers No. 15, Royal Institute of Chemistry.

The Chemical Society, London (1961) *Handbook for Chemical Society authors*, special publication No. 14.

Cahn, R. S. (1968) *An introduction to chemical nomenclature*, 3rd edition, Butterworths.

The Association for Science Education (1969) *S.I. units, signs, symbols and abbreviations*.

The Association for Science Education (1970) *Chemical nomenclature*.

Chemistry education

Royal Institute of Chemistry, Monographs for teachers. Fifteen titles have so far appeared.

Millen, D. J., ed. (1965) *Modern chemistry and the sixth form*, Collins.

Schools Council (1966) *Science in the sixth form*, Working paper No. 4, H.M.S.O.

University of New South Wales, *Approach to chemistry*. Lectures and work-shop reports of summer schools for chemistry teachers at the University of New South Wales from 1962 onwards.

Royal Institute of Chemistry, Lectures, summaries and discussion notes of Summer Schools in Chemical Education 1965, 1967, and 1968, a limited supply available on application to the Royal Institute of Chemistry.

Data Books

Nuffield Advanced Science *Book of Data*, Penguin Books.

Nuffield O-level Chemistry Project (1968) *Book of Data*, Longman/Penguin Books.

Chemical Rubber Company (1967) *Handbook of chemistry and physics*, 48th edition, Chemical Rubber Company, Cleveland, Ohio; Blackwell, Oxford.

Aylward, G. H. and Findlay, T. J. V. (1966) *Chemical data Book*, 2nd edition, John Wiley, Australasia Pty. Ltd.

National Bureau of Standards of America *Technical note* 270/1 & 2 (1965). This publication largely supersedes

Circular of the National Bureau of Standards 500 *Selected values of chemical thermodynamic properties* (1952).

Electrochemistry

Latimer, W. M. (1952) *The oxidation states of the elements and their potentials in aqueous solutions*, 2nd edition, Prentice-Hall.

Robinson, R. A. and Stokes, R. H. (1968) *Electrolyte solutions*, 2nd edition (revised), Butterworths.

Davies, C. W. (1968) *Principles of electrolysis*, Monograph for teachers No. 1, 2nd edition, Royal Institute of Chemistry.

Sharpe, A. G. (1968) *Principles of oxidation and reduction*, Monograph for teachers No. 2, 2nd edition, Royal Institute of Chemistry.

Examinations and Assessment

University of London (1967) Nuffield O-level Chemistry Examinations, Teacher's Booklet and Candidate's Booklet, obtainable from the School Examinations Department.

Bloom, B. S., ed. (1956) *Taxonomy of educational objectives*, Volume I, Longman.

Kerr, J. F. (1963) *Practical work in school science*, Leicester University Press.

Nedelsky, L. (1965) *Science teaching and testing*, Harcourt, Brace, and World.

Schools Council (1967) *Examinations Bulletin* No. 15, Teachers' experience of school-based examining (English and physics), HMSO.

Schools Council (1964) *Examinations Bulletin* No. 3, An introduction to some techniques of examining, HMSO.

History of Chemistry

Partington, J. R. (1961–1964) *A history of chemistry*, Volumes 2 to 4, Macmillan.

Knight, D. M., ed. (1968) *Classical scientific papers : chemistry*, Mills and Boon.

Leicester, H. M. and Klickstein, H. S. (1952) *A source book in chemistry*, Alembic Club Reprints (1932), Oliver and Boyd.

Inorganic Chemistry

Sidgwick, N. V. (1962) *The chemical elements and their compounds*, 2 volumes, Oxford University Press.

Cotton, F. A. and Wilkinson, G. (1967) *Advanced inorganic chemistry, a comprehensive text*, 2nd edition, Interscience.

Phillips, C. S. G. and Williams, R. J. P. (1965) *Inorganic chemistry*, Volumes I and II, Oxford University Press.

Palmer, W. G. (1965) *Experimental inorganic chemistry*, Cambridge University Press.

Moeller, T. (1952) *Inorganic chemistry, an advanced textbook*, Wiley.

Harvey, K. B. and Porter, G. B. (1963) *Introduction to physical inorganic chemistry*, Addison-Wesley.

Holliday, A. K. and Massey, A. G. (1965) *Inorganic chemistry in non-aqueous solvents*, Pergamon.

Lagowski, J. J., ed. (1966) *The chemistry of non-aqueous solvents*, Volume I, Academic Press, New York.

Day, F. H. (1963) *The chemical elements in nature*, Harrap.

Journals

School Science Review, published quarterly by John Murray for the Association for Science Education.

Education in Chemistry, published bi-monthly by the Royal Institute of Chemistry.

Journal of Chemical Education, published monthly by the Division of Chemical Education of the American Chemical Society.

Quarterly Reviews of the Chemical Society.

R.I.C. Reviews, published bi-annually by the Royal Institute of Chemistry.

Scientific American, published monthly. Selected articles on topics of interest can be obtained as *Scientific American Offprints*.

Physics Education, published bi-monthly, by the Institute of Physics and the Physical Society.

Journal of Biological Education, published quarterly by Academic Press for the Institute of Biology.

Learned societies

The Association for Science Education, College Lane, Hatfield, Herts.

The Chemical Society of London, Burlington House, London W1.

The Royal Institute of Chemistry, 30 Russell Square, London WC1.

The Society of Chemical Industry, 14 Belgrave Square, London SW1.

Other helpful information is to be found in:

Wilson, R. W. (1968) *Useful addresses for science teachers*, Edward Arnold.

Models and model-making

Nuffield O-level Chemistry Project (1968) *Handbook for Teachers*, Chapter 14, 'Models and their uses'.

Sanderson, R. T. (1962) *Teaching chemistry with models*, Van Nostrand.

Bassow, H. (1968) *Construction and use of atomic and molecular models*, Pergamon.

Organic chemistry

Sykes, P. (1967) *A guidebook to mechanism in organic chemistry*, 2nd edition, Longman.

Whitfield, R. C. (1967) *A guide to understanding basic organic reactions*, Longman.

Mann, F. G. and Saunders, B. C. (1964) *Practical organic chemistry*, 4th edition, Longman.

Ingold, C. K. (1953) *Structure and mechanism in organic chemistry*, Bell.

Clark, N. G. (1964) *Modern organic chemistry*, Oxford University Press.
Fieser, L. F. and Fieser, M. (1961) *Advanced organic chemistry*, Reinhold.
Helmkamp, G. K. and Johnson, H. W. (1964) *Selected experiments in organic chemistry*, W. H. Freeman.
Finar, I. L. (1963) *Organic chemistry* (2 volumes), 4th edition, Longman.
Openshaw, H. T. (1965) *A laboratory manual of qualitative organic analysis*, 3rd edition, Cambridge University Press.

Other science curriculum development projects

Nuffield Advanced Science: Biology, Physical Science, Physics.
Nuffield O-level Chemistry.
Nuffield O-level Physics.
Nuffield O-level Biology.
Chemical Education Material Study (USA).
Chemical Bond Approach (USA).
Approach to Chemistry (1965). Lectures and workshop reports of summer schools for Chemistry teachers at the university of New South Wales from 1962 onwards.
Details of many other curriculum projects are to be found in the Reports of the International Clearinghouse on Science and Mathematics Curricular Developments, 1962 onwards (latest available at the time of going to press is the Sixth Report, 1968), edited under the direction of J. David Lockard and published jointly for the American Association for the Advancement of Science and the Science Teaching Centre, University of Maryland, by the University of Maryland, USA.

Structure and Bonding

Wells, A. F. (1962) *Structural inorganic chemistry*, 3rd edition, Oxford University Press.
Chemical Society, London (1958) *Interatomic Distances*, Special publication No. 11.
Chemical Society, London (1965) *Interatomic Distances Supplement*, Special Publication No. 18.
Cottrell, T. L. (1958) *The strengths of chemical bonds*, 2nd edition, Butterworths.
Coulson, C. A. (1961) *Valence*, 2nd edition, Oxford University Press.
Spice, J. E. (1966) *Chemical binding and structure*, Pergamon.
Cartmell, E. and Fowles, G. W. A. (1966) *Valency and molecular structure*, 3rd edition, Butterworths.
Linnett, J. W. (1964) *The electronic structure of molecules: a new approach*, Methuen.

Herzberg, G. (1944) *Atomic spectra and atomic structure*, Dover, New York.
Hochstrasser, R. M. (1964) *Behavior of electrons in atoms*, Benjamin.
Wheatley, P. J. (1959) *The determination of molecular structure*, Oxford University Press.
Holden, A. and Singer, P. (1964) *Crystals and crystal growing*, Heinemann.

Tested experiments and demonstrations
Nuffield O-level Chemistry Project (1967) *Collected Experiments*, Longman/Penguin Books.
Fowles, G. (1963) *Lecture experiments in chemistry*, 6th edition, Bell.
Association for Science Education (1964) *Tested experiments for use with Chemistry for Grammar Schools*, A.S.E. (John Murray).
American Chemical Society (1966) *Tested demonstrations in chemistry*, compiled by Alyea, H. N. and Dutton, F. B. 6th edition.

Index

References to specific salts are indexed under the name of the appropriate cation; references to substituted organic compounds are indexed under the name of the parent compound.

Acknowledgements

The Nuffield Foundation Science Teaching Project puts on record its warmest thanks to the following whose advice, assistance and support were much valued by the Headquarters Team of the Advanced Chemistry Section. They would also like to acknowledge here with pleasure and gratitude a generous financial contribution from Charter Consolidated Limited without which production of *The Chemist in Action* would have been impossible.

The Joint Committee for Physical Sciences

Professor Sir Nevill Mott, FRS (Chairman)
Professor Sir Ronald Nyholm, FRS (Vice-Chairman)
Professor J. T. Allanson
Dr P. J. Black
N. Booth
Professor C. C. Butler, FRS
E. H. Coulson
D. C. Firth
Dr J. R. Garrood
Dr A. D. C. Grassie
Professor H. F. Halliwell

Miss S. J. Hill
Miss D. M. Kett
Professor K. W. Keohane
Professor J. Lewis
J. L. Lewis
A. J. Mee
Professor D. J. Millen
J. M. Ogborn
E. Shire
Dr J. E. Spice
Dr P. Sykes
E. W. Tapper
C. L. Williams

Working Parties
Main Course

A. Ashman
Professor P. G. Ashmore
M. G. Brown
Professor A. D. Buckingham
Dr B. E. Dawson
Professor E. A. V. Ebsworth
Professor V. Gold
Dr A. D. C. Grassie
Professor H. F. Halliwell
A. Jackson

Dr T. J. King
Professor H. C. Longuet-Higgins, FRS
Dr R. S. Lowrie
Dr R. B. Moyes
T. A. G. Silk
M. Smith
Dr P. Sykes
Professor J. M. Tedder
Professor A. F. Trotman-Dickinson

Special Studies

J. P. Allen

Dr C. B. Amphlett

J. M. V. Blanshard

Dr G. S. Boyd

A. A. Brown

Dr J. I. Bullock

J. A. Charles

Dr J. P. Chilton

P. S. Coles

Professor E. M. Crook

Dr R. J. Dalton

Dr R. W. Davidge

Miss A. A. Free

R. Grace

P. Handisyde

D. W. Harding

Dr S. Harper

E. K. Hayton

G. C. Hill

Professor A. T. James

Dr P. L. James

K. S. Jepson

P. J. Kelly

J. A. Kent

Dr J. A. Knight

Dr R. A. Lawrie

Dr J. W. Martin

Dr B. E. Mayer-Abraham

Dr A. M. Mearns

Dr H. G. Muller

K. E. Peet

J. Pilot

Dr B. R. W. Pinsent

D. L. Rowlands

D. K. Rowley

Professor J. E. Salmon

G. C. Smith

Dr G. Stainsby

R. J. Sutcliffe

Professor J. D. Thornton

Professor A. G. Ward

K. Watson

Dr F. C. Webb

Dr E. D. Wills

Others who have given advice and assistance

Professor C. C. Addison, FRS

Dr Barbara E. C. Banks

W. R. H. Batty

Professor J. C. Bevington

A. A. Bishop

Dr N. G. Clark

Miss A. Gittins

P. Gradwell

Professor E. A. Guggenheim, FRS

Dr D. Halliday

Dr J. H. Harker

Dr G. J. Howard

J. Lamb

C. N. Lammiman

Dr D. A. Lewis

Dame Kathleen Lonsdale, FRS

Miss E. W. McCreath

Professor M. L. McGlashan

Dr J. Manning

Dr A. Miller

Professor D. C. Phillips, FRS

Professor J. C. Robb

R. A. Ross

Professor P. N. Rowe

Dr J. Saul

A. E. Somerfield

Dr M. H. B. Stiddard

M. Taylor

R. J. Taylor

P. Thompson

Dr C. Tittle

J. R. White

A. S. Willmott

Chemistry teachers who were involved in teaching the draft material
in the Trials' Schools

Dr A. W. Bamford
D. J. Boyne
Mrs M. Bristow
C. A. Clark
M. G. Coldham
R. S. Cowie
D. J. Defoe
Dr F. A. Downing
C. D. Evans
Dr H. J. C. Ferguson
L. Ferguson
K. Fraser
Miss A. A. Free
A. J. Furse
A. Gill
N. N. Gilpin
T. Given
D. C. Hobson
Dr G. Huse
D. G. Hutt
Mrs M. P. Jarman
Miss A. M. Lewis
Miss P. G. Lineton
Dr N. H. Lumb
M. R. Mander
D. H. Mansfield
Dr R. T. B. McClean
G. McClelland

D. McCormick
S. J. McGuffin
J. M. McNevin
K. D. Millican
S. Mitchell
D. W. Muffett
Mrs H. G. Naylor
C. Nicholls
Miss J. Peck
R. J. Porter
A. G. Rayden
H. Read
Mrs M. Reed
Dr K. Reid
G. P. Rendle
C. J. V. Rintoul
M. J. W. Rogers
R. Sprawling
D. M. Stebbens
D. A. Stephens
R. J. Stephenson
R. J. Sutcliffe
Mrs G. I. Tait
W. R. Towns
N. R. Waite
A. P. Ward
S. C. White
Miss M. Wilson

Schools which took part in the trials

From September, 1966
Blyth Grammar School
Braintree County High School
Deeside Senior High School, Flints.
Edinburgh Academy
Huddersfield New College
King's College School, Wimbledon
Methodist College, Belfast
Nottingham High School for Girls
Pontefract Girls' High School, Yorks
Prendergast Grammar School,
  Catford, London SE6
Royal Grammar School, High
  Wycombe
Westminster School

Additional schools from September, 1967
Bradford Girls' Grammar School
Clifton College, Bristol
Grosvenor High School, Belfast
King Edward VI Grammar School,
  Aston, Birmingham
Maidstone Grammar School
Malvern College
Newport High School
Oundle School
Rossall School, Fleetwood, Lancs.